Climate Change in Africa

Michael Addaney · D B Jarbandhan ·
William Kwadwo Dumenu
Editors

Climate Change in Africa

Adaptation, Resilience, and Policy Innovations

palgrave
macmillan

Editors
Michael Addaney
Department of Land Management
University of Energy and Natural Resources
Sunyani, Ghana

D B Jarbandhan
Centre for Public Management and Governance
University of Johannesburg
Auckland Park, South Africa

William Kwadwo Dumenu
Kumasi, Ghana

ISBN 978-3-031-30049-3   ISBN 978-3-031-30050-9 (eBook)
https://doi.org/10.1007/978-3-031-30050-9

Cover credit: Kypros

This Palgrave Macmillan imprint is published by the registered company Springer Nature Switzerland AG
The registered company address is: Gewerbestrasse 11, 6330 Cham, Switzerland

Preface

This book, *Climate Change in Africa: Adaptation, Resilience, and Policy Innovations,* offers readers a comprehensive understanding of the salient content and key developments in climate change adaptation, resilience, and policy action at the regional level in Africa with selected sub-regional and local case studies. Climate change is recognized as one of the most critical and topical issues facing the world over the past decades, but has become more intense in the twenty-first century. The associated impacts of climate change such as warming, sea-level rise, extreme changing weather patterns and events, pressure on water and food security, human health risks, and impacts on wildlife and ecosystems are considered as major existential threats to developing regions and societies. African countries catalog some of the greatest damaging impacts of climate change even though they account for less than 5 percent of the global total greenhouse gas emissions. The current and projected climate change poses a set of fundamental challenges to local economies and livelihoods across the African continent, contributing to harsh physical and environmental conditions that are inducing human mobility, health crises, disasters, weakened social and cultural resilience, worsening energy generation and distribution, and rising public sector debts. The heterogeneity of African countries also presents special challenges for localized climate projections, mitigation action, and adaptation strategies. Effective mitigation and adaptation measures and policies to address climate change impacts and vulnerabilities in Africa should, therefore, acknowledge their unique economic, social, environmental, and cultural characteristics.

The chapters chronicle the public, municipal, and communal policy responses to current and potential climate change impacts that will affect critical and priority sectors within and across countries now and in the coming decades. Thus, the chapters in this book cover governance and policy responses to climate change, emphasizing continental governance and policy responses, national governance and policy responses (what selected countries in Africa are doing), and local or community policy responses (what some selected major African cities and communities are doing). With chapters written in an accessible style, the book carries timely contributions within the contemporary state of the art of climate change adaptation, resilience, and policy research. It will thus be useful for academics and postgraduate students within disciplines such as geography, urban planning and policy design, political science, sociology, anthropology, and public law. The chapters have undergone rigorous peer review process to guarantee that readers gain a discerning insight into the nuances of climate change adaptation, resilience, and policy-making in Africa. The editors and contributors include both established and emerging scholars and practitioners. Both the focus of the book and the voices emerging from the various chapters make this a very timely publication.

Sunyani, Ghana — Michael Addaney
Johanneburg, Ghana — D B Jarbandhan
Kleve, Germany — William Kwadwo Dumenu

Acknowledgments

This book is culmination of many years of work that would not have been possible without the effort and patience of all the contributing authors—thank you to everyone involved for your commitment to and participation in the project. We also wish to express our gratitude to the following people who provided encouragement, contributed to and supported the ideas, design, and overall development of the book—the editorial team at Palgrave MacMillan, Rachael Ballard (commissioning editor), and Cecile Schuetze-Gaukel (project coordinator) as well as the copyeditors and the anonymous reviewers were very supportive. They assisted in shaping the concept of the book and refined the final output to make sure that the publication was on schedule. Our sincere appreciation goes to all who assisted and supported this vision to completion, especially to colleagues who assisted in reviewing some of the chapters. We are grateful for the time you committed to support this project.

Contents

Notes on Contributors

Michael Addaney is a lecturer in land use planning and environmental law at the Department of Land Management of University of Energy and Natural Resources, Ghana and senior research associate at the Centre for Public Management and Governance at the University of Johannesburg, South Africa. He is also a research fellow of the Earth System Governance Project at Utrecht University, the Netherlands. He is an environmental social scientist whose expertise and current research interests are in the multifaceted and embedded relationships between humans and the environment, whether facilitated by institutions or by local organizations/communities, and the effects of this on public policy and planning processes and outcomes, particularly in relation to notions of rights, justice, and equity. He holds a Ph.D. in environment and natural resources law from the Wuhan University, China and B.Sc. Development Planning from the Kwame Nkrumah University of Science and Technology, Ghana.

Olubunmi A. Afinowi is a lecturer and researcher at the Department of Private and Property Law, Faculty of Law, University of Lagos, Nigeria. Her areas of research include climate change law and governance, sustainable development law, and international environmental law. Her Ph.D. research was on the role of planning legislation to climate change adaptation in Nigeria and South Africa. Other areas of research interest include the conservation of natural resources in Africa and coastal governance in Nigeria. She teaches several courses, including International Environmental Law, Housing Law, and Planning Law at both undergraduate and postgraduate levels.

Timothy Amang-bey Akanpabadai is a practice-based development planner with 12 years' experience and interests in local economic development, climate change and adaptive capacities, and food security. He works with World Vision Ghana as Regional Operations Manager at its Northern Regions Operations.

Lenin Tinashe Chisaira is an environmental lawyer and researcher based in Harare, Zimbabwe. He is the founder and director of Advocates4Earth, a non-profit, public interest environmental law, climate, and wildlife justice organization. He holds an LLM International Human Rights Law and Public Policy (University College Cork, Ireland); an LLM Economic Regulation, PGCert in Environmental and Natural Resources Law (University of London); and an LLB Hons (University of Zimbabwe.). One of his latest publications is a book chapter on "Human Rights and Energy in a Neoliberal Southern Africa" in Addaney M., Oluborode Jegede A. (eds) *Human Rights and the Environment under African Union Law* (Palgrave Macmillan, Cham. 2020).

Stephen Kofi Diko is an assistant professor at the University of Memphis, USA. He holds a Ph.D. in Regional Development Planning from the University of Cincinnati, Ohio. His research examines sustainable urban development and policy from the aspects of climate change, green spaces, flooding, informality, and local economic development.

Felix Dube is a postdoctoral fellow in the Department of Public, Constitutional, and International Law, College of Law, University of South Africa. He completed a Bachelor of Laws from the University of Venda, after which he successfully pursued a Master of Laws and Doctor of Laws from the North-West University. During his doctoral studies, he taught constitutional law and legal history at North-West University. His research interests are in constitutional hermeneutics, human rights theory, and selected aspects of public and private law that link to constitutional law.

Marus Gbomagba is Ph.D. candidate in International Environmental Law at the Research Institute of Environmental Law of Wuhan University, China. He is an assistant lecturer at the Faculty of Law of the University of Abomey-Calavi, Benin. He is also a front office intern at the UNDP-Regional Bureau Africa (RBA) and the UNSSC Circular Economy Cohort 2020/2021 as well as the United Nations International Law Program Fellow.

D B Jarbandhan is associate professor, former Head of Department of Public Management and Governance and is currently the director of the Centre for Public Management and Governance in the College of Business and Economics, University of Johannesburg, South Africa. His academic interest focuses on, inter alia, public sector human resource management, future-fit leadership and development, and good governance. He is an experienced and registered educator. He served on the Executive Committee of the Association and Schools of Southern African Departments of Public Administration and Management (ASSADPAM), and he had the privilege to serve as the Chief Editor of Administratio Publica, a DoHET-accredited journal (2015–2019). He now serves on the Editorial Board of Administratio Publica. He has also supervised honors, masters, and doctoral students to completion and published widely in accredited journals. He currently serves on the expert panel of Innovating Governance, an international NGO, which is based in Vienna, Austria.

Naomi Kumi is a lecturer and researcher at the Department of Atmospheric and Climate Science of the University of Energy and Natural Resources (UENR), Sunyani, Ghana. She also doubles as the examination officer and the postgraduate studies coordinator of the Department. She is also a fellow of the AIMS-NEI Fellowship Program for Women in Climate Science (WiCCS) and founder of the Association for Women in Climate Science (AWiCS) in UENR.

William Kwadwo Dumenu is a research scientist at the Forestry Research Institute of Ghana and a research associate and lecturer at the Sustainable Food System Institute, Rhine-Waal University of Applied Science, Germany. He is also currently pursuing Ph.D. in forest sciences at the University of Goettingen, Germany. His research focuses on forest governance, economics, and sustainability of biodiversity and ecosystem services, innovations and entrepreneurial activities involving forest products, and valorization of non-timber forest products for rural livelihood and sustainable economic development. Over the years, he has led and participated in research projects and consultancies on equitable distribution of REDD+ costs and benefits, climate change impact on forest livelihoods, biophysical and economic valuation of ecosystem services for organizations, such as International Union for Conservation of Nature (IUCN), World Agroforestry (ICRAF), World Wildlife Fund (WWF), United Nations Development Programme (UNDP), Government of Ghana, and the German Federal Ministry of Education and Research.

Aubrey Manthwa is the director of the School of Law in the College of Law, University of South Africa (UNISA). He obtained his LLB, LLM, and LLD at UNISA. He has also served as the Head of the Department of Public, Constitutional, and International Law (PCIL) in the College of Law. Presently, he serves in various College of Law and institutional portfolios and committees, such as the Senate and ICT steering Committee at UNISA. Prior to assuming the roles of Head of Department and Director, he was a senior lecturer and taught various modules in both private and public law at undergraduate and postgraduate levels. His research interests include African law, constitutional law, family law, and administrative law. He is particularly interested in the role of Western constitutions in shaping African legal systems. In his research, he is inspired by the Africanization struggles of South Africa and its counterparts on the continent.

Zenebe Mekonnen is senior researcher and director of climate Science Research directorate with research focus of climate change mitigation and adaptation as well as climate change modeling and risk management at Ethiopian Environment and Forest Research Institute (Currently restructured as Ethiopian Forestry Development). He has Ph.D. in Climate Change and Bioenergy Development (Climate change mitigation and adaptation) from Hawassa University, Ethiopia. He has been also researcher at Ethiopian Institute of Agricultural Research with key research duties focused on the socioeconomics of non-timber forest products, agroforestry, medicinal plants, and wood fuel energy. He had worked in government and non-government institutions in different technical and expert positions. He has over thirty papers published in peer-reviewed journals, chapters in books, and conference and/or workshop proceedings.

Seth Opoku Mensah is a doctoral student at the Institute for Sustainable Futures, University of Technology Sydney, Australia. He is an interdisciplinary development planner with research interest in human-environment interactions with particular focus on climate change and livelihood dynamics, climate change adaptation, climate smart agriculture, sustainable land management practices, and natural resource management. He is a member of Ghana Institute of Planners (GIP), Commonwealth Association of Planners (CAP), and the Ghana Monitoring and Evaluation Forum (GMEF).

Carol Chi Ngang is a senior lecturer at the Department of Public Law, National University of Lesotho, research fellow at the Free State Centre for Human Rights, University of the Free State, and a guest lecturer at the Centre for Human Rights, University of Pretoria. He is the UNDP Human Rights Research Chair for Lesotho and the South African National Research Foundation Category C2-rated researcher. He has been a visiting researcher to the Van Vollenhoven Institute, Leiden University in 2022, Law and Development Research Group, University of Antwerp in 2016, and Cegla Centre for Interdisciplinary Research in Law, University of Tel-Aviv in 2015. His research interest combines human rights and development with a niche focus on the right to development and socio-economic and cultural rights wherein, he has published extensively including books, book chapters, and journal articles. He is a member of the Law and Development Research Network (LDRN) and Law and Society Association (LSA), associate editor of the *Lesotho Law Journal (LLJ)* and advisory board member of the *International Journal of Ethiopia Legal Studies* (*IJELS*). He has worked previously within the NGO sector in Cameroon and South Africa.

Rachel Yeboah Nketiah is an assistant research fellow at the Centre for Climate Change and Gender Studies of the University for Energy and Natural Resources, Ghana. She is a social scientist with expertise in community development, advocacy with particular interest in women achieving justice and equity in terms of development. She holds a B.A. degree in Planning and Rural Development and MPhil in Development Studies from the University of Development Studies, Tamale, Ghana.

Nelson Nyabanyi N-yanbini is an assistant lecturer at the Department of Planning in Simon Diedong Dombo University of Business and Integrated Development Studies, Ghana. He holds an MPhil in Planning and a Bachelor of Science in Human Settlement Planning, both from the Kwame Nkrumah University of Science and Technology (KNUST). He previously worked as a Project Coordinator in Family Support Lifeline (NGO) and as a teaching and research assistant in at the Department of Planning, KNUST. His areas of research interests are climate change adaptation, urban planning, urban design and sustainability, and land governance.

Seth Asare Okyere is a development planner and a visiting assistant professor at the College of Architecture, Planning, and Landscape Architecture, University of Arizona (USA). He holds a Ph.D. in Engineering with emphasis on urban development planning from Osaka University. His research interests and experiences focus on interdisciplinary themes around development planning, urbanism, and sustainable development in global south cities.

Caiphas Brewsters Soyapi is researcher at the North-West University, South Africa. His research interests currently focus primarily on environmental constitutionalism within Africa. Within this space, he broadly explores rights-based environmental protection, the place of international environmental law principles in African courts, and African judicial environmentalism.

Habib Sani Usman is an instructor at the School of Law, American University of Nigeria. He attended University of Maiduguri for a Bachelor's degree in Common Law and Sharia (LLB and Sharia). Thereafter, he was called to the Nigerian Bar as Barrister and Solicitor of the Supreme Court of Nigeria. He was in private practice between 2006 and 2010 until he took up a teaching appointment with Modibbo Adama University of Technology, Yola as assistant lecturer. He holds a Master of Laws degree (LLM) in Law and Technology from Tilburg University, the Netherlands. He is currently conducting research on the development of agriculture-focused adaptation laws in Kenya, Nigeria, and South Africa at the North-West University, Potchefstroom, South Africa.

Abbreviations and Acronyms

ACHPR	African Charter on Human and People's Rights
ADF	African Development Forum
AfDB	African Development Bank
AMCEN	African Ministerial Conference on Environment
AU	African Union
AUC	African Union Commission
C2ES	Centre for Climate and Energy Solutions
CAADP	Comprehensive Africa Agriculture Development Programme
CAHOSCC	Committee of African Heads of State and Government on Climate Change
CBO	Community-based Organisations
CCA	Climate Change Act of 2021
CCTCCP	City of Cape Town Climate Change Policy of 2017
CDM	Clean Development Mechanism
CIFOR	Centre for International Forestry Research
COMESA	Common Market for East and Sub-Saharan Africa
COP	Conference of Parties
COP26	26th Conference of Parties
CRGE	Climate Resilient Green Economy
CSA	Climate-smart Agriculture
CSO	Civil Society Organisations
DoA	Department of Agriculture
DRMA	Disaster Risk Management Act No 57 2002
EAC	East African Community
ECOWAS	Economic Community of West African States
EHoA	East and the Horn of Africa

ENRSIP	Environment Natural Resources Sector Investment Plan
EPA	Environmental Protection Agency
ERPA	Emission Reductions Payment Agreement
FAO	Food and Agriculture Organizations of the United Nations
FCPF	Forest Carbon Partnership Facility
FDRE	Federal Democratic Republic of Ethiopia
FEWNET	Famine Early Warning Networks
FGDs	Focus Group Discussions
FRL	Forest Reference Level
GHG	Greenhouse gas
GMA	Ghana Meteorological Agency
GOU	Government of Uganda
GSS	Ghana Statistical Service
HFIAS	Household Food Insecurity Access Scale
HIV/AIDS	Human Immunodeficiency Virus/Acquired Immunodeficiency Syndrome
HOA	Horn of Africa
ICCPR	International Covenant on Civil and Political Rights
ICESCR	International Covenant on Economic, Social, and Cultural Rights
ICRAF	International Centre for Research in Agroforestry
IDP	Integrated Development Plan
IGAD	Intergovernmental Authority on Development
IPCC	Intergovernmental Panel on Climate Change
KIIs	Key Informant Interviews
LAS-CCAS	Lagos State Climate Change Adaptation Strategy
LC/LUC	Land Cover/Land Use Change
LDCs	Least Developed Countries
LULUCF	Land Use, Land Use Change, and Forestry
MAAIF	Ministry of Agriculture, Animal Industry and Fisheries
MDGs	Millennium Development Goals
MEFCC	Ministry of Environment, Forest and Climate Change
MMT	Million Metric Tonnes
MoFA	Ministry of Food and Agriculture
MOP	Meeting of Parties
MRV	Measurement, Reporting, and Verification
NAP	National Adaptation Programme
NASPA-CCN	National Adaptation Strategy and Plan of Action on Climate Change for Nigeria
NCCAP	National Climate Change Action Plan
NCCAS	National Climate Change Adaptation Strategy
NCCC	National Climate Change Council
NCCP	National Climate Change Policy

NCCR	National Climate Change Response
NCCRP	National Climate Change Response Policy White Paper
NCCRWP	National Climate Change Response White Paper
NDC	Nationally Determined Contributions
NGO	Non-Governmental Organisation
NGOs	Non-Governmental Organisations
NRS	National REDD+ Strategy
OAU	Organisation of African Unity
ODK	Open Data Kit
OECD	Organisation for Economic Cooperation and Development
PAP	Pan African Parliament
PFM	Participatory forest management
PoU	Prevalence of Undernutrition
RDI	Rural Development Institute
REC	Regional Economic Communities
REDD	Reduced Emissions from Deforestation and Forest Degradation
REDD+	Reducing Emission from Deforestation and Degradation
R-PIN	Readiness Plan Program Idea Note
R-PP	Readiness Preparation Proposal
S4T	Savings for Transformation
SDGs	Sustainable Development Goals
SIS	Safeguard Information System
tCO2e	tonnes of Carbon dioxide equivalents
TWAIL	Third World Approaches to International Law
UDHR	Universal Declaration of Human Rights
UN	United Nations
UNCCD	United Nations Convention to Combat Desertification
UNCCP	Uganda National Climate Change Policy
UNCTAD	United Nations Conference on Trade and Development
UNDP	United Nations Development Programme
UNDPA	United Nations Department of Political Affairs
UNECA	United Nations Economic Commission for Africa
UNEP	United Nations Environment Programme
UNFCCC	United Nations Framework Convention on Climate Change
UNHCR	United Nations High Commissioner for Refugees
US	United States
USAID	United States Agency for International Development
WBG	World Bank Group
WBISPP	Woody Biomass Inventory and Strategic Project
WCCCRS	Western Cape Climate Change Response Strategy 2014
WFP	World Food Programme
WHO	World Health Organisation
WRI	World Resources Institute
WVA	World Vision Australia
WVE	World Vision Ethiopia

List of Figures

Chapter 10

List of Tables

Chapter 10

CHAPTER 1

Introduction to Climate Change, Adaptation Policy and Resilient Development in Africa

Michael Addaney, William Kwadwo Dumenu, and D B Jarbandhan

1 Climate Change in Africa: Overview of Impacts, Adaptation and Resilience Policy Responses

Climate change is unambiguously recognised as human-induced and poses significant threat to human survival and socio-ecological systems (Intergovernmental Panel on Climate Change [IPCC], 2021). Recent increases in global temperatures are unprecedented and projected to accelerate. Even extreme limitation of greenhouse gas emissions can only

M. Addaney (✉)
Department of Land Management, University of Energy and Natural Resources, Sunyani, Ghana
e-mail: Michael.Addaney@uenr.edu.gh

W. Kwadwo Dumenu
Forestry Research Institute of Ghana, Ejisu, Ghana
e-mail: wdumenu@csir-forig.org.gh

Rhine-Waal University of Applied Science, Kleve, Germany

M. Addaney et al. (eds.), *Climate Change in Africa*,
https://doi.org/10.1007/978-3-031-30050-9_1

slow the pace of temperature increases, given that past emissions remain in the atmosphere (IPCC, 2018). The 0.7 °C rise in global temperatures over the past 30 years (or 1 °C over the past 50 years) is significantly higher than in any equivalent period during the last 10,000 years (Marcott et al., 2013). Natural disasters have been occurring but there is strong evidence that increasing temperatures and changes in precipitation levels contribute to more frequent droughts, desertification, rising sea levels and higher vapour pressure—most of which drives more frequent floods and storms (IPCC, 2018). Scientific evidence indicates that the impacts of climate change is global and every continent and country is affected (IPCC, 2021). For example, according to the IPCC (2021), human influence has likely increased the chance of compound extreme events since the 1950s including increases in the frequency of concurrent heatwaves and droughts in some regions across continents and compound flooding in some locations. However, the impacts of climate change affect developing countries and regions differently due to their level of development, unique geographies and traditional development challenges (Addaney & Cobbinah, 2019; Schellnhuber, 2013).

Extreme climate events such as increasing temperatures, rising sea levels and unpredictable rainfall patterns are increasing the occurrence and intensity of natural disasters and are markedly transforming the African continent's geography (IPCC, 2018). For instance, recent natural disasters such as the devastating cyclones Idai and Kenneth; locust invasion in eastern Africa and droughts in southern and eastern Africa as well as rapidly expanding Saharan desert are threatening the lives of millions and contributing to conflicts and mass migration (Rigaud et al., 2018). The outcomes of these events leave millions in peril, injured, homeless or food insecure, and inflict grave and costly economic damage. The projected impact of extreme weather events in sub-Saharan Africa is higher and extensive than other regions in the world due to the continent's lower resilience and adaptive capacity. For instance, extreme climate events in Africa are responsible for an estimated 1,000 deaths with some 13 million people being seriously affected (injured, left homeless, food insecure or

D B Jarbandhan
Centre for Public Management and Governance, University of Johannesburg, Johannesburg, South Africa
e-mail: vainj@uj.ac.za

lacking water and sanitation) annually (International Monetary Fund, 2016).

The impact of climate change on rainfall patterns is more multifaceted in Africa as western and southern African countries, including some of sub-Saharan Africa's driest locations, are experiencing marked precipitation declines (Nyiwul, 2019). Other countries such as Madagascar, Malawi, South Africa and Zimbabwe are having some areas rapidly drying out while others are experiencing immense rainfall increases. On the other hand, the Sahel and south-eastern Africa are severely affected by droughts induced by prolonged heat and dryness and having a serious toll on people's lives and livelihoods as it is eroding the progress achieved in the last three decades' in raising life expectancy, malnutrition and reducing infant mortality (African Union, 2020). Also, Comoros, Madagascar, Malawi and Mozambique are particularly vulnerable to tropical cyclones from the Indian Ocean while Guinea Bissau and Sierra Leone are vulnerable to storms from the Atlantic Ocean (African Union, 2020; IMF, 2021). Thus, large coastal cities (such as Abidjan, Accra, Dakar, Dar es Salaam and Lagos) are vulnerable to floods from rising sea levels. Overall, the rate of sea-level rise around Africa is higher than the global mean (IPCC, 2014). In 2021, many extreme and high-impact events were reported across Africa including heavy rainfall, floods, land-falling tropical cyclones, droughts, heatwaves, wildfires and sandstorms (IMF, 2021). These events led to disastrous impacts on socioeconomic sectors.

The increasing threat and impact of climate change have provoked sustained call and search for effective mitigation and adaption measures, and the building of climate resilience. Countries and States party to the UNFCCC have pursued this by developing National Adaptation Plans, and local adaptation policies, strategies and legislation to addressing climate change impacts and risks (Averchenkova et al., 2017). Adaptation refers to processes or actions taken to adjust to the actual or expected climate change and its effects (IPCC, 2014). Adaptation debates often focus on promoting specific urgent actions before it is no longer rationally possible to adapt to, minimise or avoid harm from climate change (Dow et al., 2013). Such measures could include building sea walls to protect people against sea-level rise, installing new irrigation systems to combat water scarcity or planting trees to reduce air pollution and cool urban areas (Mayer, 2018). More infrequently, adaptation may also seek to take advantage of opportunities that climate change could bring such as making new crop choices suited to the changed climate. On the other hand, resilience refers to the ability of a social or ecological system to absorb disturbances while retaining the same basic structure and ways of functioning (IPCC, 2007). It denotes the capacity of a socio-ecological system to absorb climate change-related disturbances or shocks while retaining the same basic structure and ways of functioning (Bellinson & Chu, 2019). Climate resilience is closely linked to adaptation in literature (Scheffer et al., 2015). Both resilience and adaptation focus on consequences of extreme climate events and development of preparedness and response measures based on vulnerability assessments and suitable individual actions for maintaining stability during and after such an event. Resilience is an ongoing, iterative process which recognises that vulnerability and climate risk are constantly evolving as socio-ecological systems (infrastructure and ecosystems), local communities (with specific social and cultural norms and values) and institutional structures (policies and political actors) within them develop and interact.

Taking action on adaptation has implications for resilience as in many cases it increases resilience. For instance, the implementation of large-scale and costly adaptation projects may potentially lead to a reduction in the number of available options for responding effectively to an adverse climate impact. Adaptation can therefore be short- or long-term, a single action or a holistic programme and reactive and backward-looking or proactive and forward-looking. Resilience is more associated with system-wide changes, including a range of solutions to enhance social, human, natural, physical and financial capacities; forward-looking and anticipatory planning for long-term capacity-building; and enhancing the ability to absorb shocks and recover from their impacts. Adopting resilience thinking within climate adaptation projects and policies encourages a shift from short-term, incremental, project-focused and reactive approaches towards a kind of planning that is long-term, transformative, holistic and forward-looking. For example, flood adaptation strategies that are informed by resilience thinking encompass a range of actions for reducing risks (such as analysing hazard probability, exposure and vulnerability) before flood events as well as enhancing preparation, response and recovery capacities for when a flood occurs through various environmental, financial, social and political mechanisms. Thus, adaptation and resilience are complementary, working in conjunction with each other when it comes to responding to the impacts of climate change. As such, building resilience requires robust, revolutionary and proactive decision-making based on ongoing learning and evaluation of the success or challenges of current approaches to climate adaptation. It requires integrated and multi-sectoral approaches which take into account the context-specific risk and vulnerability, including the social, economic and environmental pressures and opportunities.

2 Scope of the Book and Preliminary Conclusions

This book explores the salient issues and key developments in climate change adaptation, resilience and policy action in Africa including the social, economic and political (governance) challenges governments and local communities face in delivering remedial measures. The chapters collectively discuss efforts that governments and local communities are making to adapt to the changing climate and, more broadly, to become more resilient against the impacts of climate change such as sea-level rise, increased storm intensity and floods, food and water insecurity, droughts and extreme temperatures. In a region where an estimated 600 million people rely mainly on climate-sensitive sectors, adapting to the risks and impacts of climate change remains a daunting task for countries across the continent actively pursuing strategies or action plans to manage climate change. Furthermore, the magnitude of the threats posed by climate change requires innovative ideas, strategies and projects. The national and local case studies involving countries such as Ethiopia, Ghana, Mozambique, Nigeria, South Africa and Zimbabwe as well as regional analyses featured in this book offer practical examples and insight for the consideration of policymakers, development practitioners.

In Chapter 2, Carol Chi Ngang interrogates the extent to which the impacts of climate change may undermine the realisation of the right to development in Africa. He re-echoes the conventional argument that Africa contributes less than 5% to carbon emissions which eventually lead to climate change. Thus, he makes a case that the obligation to resolve bottlenecks to development requires careful consideration of climate change impacts and risks. In Chapter 3, Marus Gbomagba and colleagues also explore the gap that currently exists in the coordination of climate change adaptation policies and institutions in sub-Saharan Africa in providing an enabling institutional environment for enhanced climate action. They highlight the inadequacy of existing policy frameworks for climate change adaptation governance and how it undermines adaptive capacity. Moving the discussion forward, Caiphas Brewsters Soyapi and colleagues also contextualise the significance of indigenous knowledge in facilitating vulnerable communities to overcome various problems limiting adaptation action in Africa, including poor governance, poor technologies and endemic poverty. Consequently, this chapter discusses the extent to which indigenous knowledge forms part of policy and governance responses to climate change adaptation in Africa. Chapter 5 focuses on the intersectionality of climate change and gender justice in forest-dependent communities in Africa. In this chapter, Rachel Yeboah Nketiah, Naomi Kumi and Zenebe Mekonnen examine how benefit sharing could promote inclusiveness and transparency within forest fringe communities, and eventually drive the success of REDD+ initiatives as a climate change adaptation measure. The authors emphasise the difficulties that external stakeholders are facing that impede performance and stakeholder participation in the implementation of effective forest protection programmes in Africa.

The twin forces of climate and rapid urbanisation in Africa are hindering the provision of adequate infrastructure for informal settlements. Thus, climate change interventions in these areas clearly overlap with sustainable development priorities. In Chapter 6, Olubunmi Ayodele Afinowi examines the extent of collaborations and partnerships in adaptation to climate change across levels of government in Lagos (Nigeria) and South Africa (Cape Town). She argues that adaptation action in South Africa is more collaborative as well as cross-sectoral, encompasses disaster risk management and sustainable development as viable means of building resilience while, Nigeria on the other hand has not fully integrated these principles into its climate action. In Chapter 7, Lenin Tinashe Chisaira conducts a critical assessment of the legal responses to increasing climate-induced disasters by the central and local governments of Zimbabwe and Mozambique. He contends that the governments of both countries must place more efforts on civil and human rights protection as well as ensuring compliance with international climate law obligations through the adoption of suitable climate legislation and measures.

Urban greening is important for environmental, scenic, health and socioeconomic benefits. However, the governance of urban greening is dynamic and intricate despite its usefulness in climate change adaptation. In Chapter 8, Nelson Nyabanyi N-yanbini uses good governance principles in theorising and understanding the governance of urban green spaces towards climate change adaptation in Ghanaian cities. He argued that three out of the eight good governance principles (inclusivity, fairness and accountability) were declining despite their importance in urban greening. Therefore, he forcefully makes a case for the integration of international standards with local preferences in governing urban greening. In Chapter 9, Zenebe Mekonnen discusses the centrality of good forest and carbon governance focusing on legality, participation and legitimacy. He argues that realising good forest and carbon governance in Ethiopia, where tenure is not clearly defined and corruption is deep-rooted is an ambitious agenda that covers more than the institutional architecture for administering funds and monitoring results. He concludes that land and forest tenure is a fundamental issue of concern for future carbon use policies and measures since addressing tenure can strengthen forest governance and improve prospects for forest conservation. In Chapter 10, Seth Opoku Mensah and colleagues use the Talensi district as a case study to examine climate variability and household food insecurity in Ghana. They argue that due to the agro-based livelihoods, subsistence nature and extensive reliance on rain-fed agriculture of the district, poor crop yield as a result of climate variability manifested in erratic rainfall pattern, drought and land degradation will put many households at risk of food insecurity. Given the growing appreciation of approaches that engage environmental law from the lens of indigenous philosophical foundations and knowledge systems, in Chapter 12, Felix Dube and Aubrey Manthwa argue that indigenous values that command a strong presence in South Africa's environmental policy and regulatory approaches are the ideal premises from which to secure ecologically sustainable development. In this regard, they identify *Batho Pele*, the indigenous value of putting people first in all decisions that affect them as an essential component of balancing the competing interests of protecting ecological integrity and achieving economic and social development.

References

Addaney, M., & Cobbinah, P. B. (2019). Climate change, urban planning and sustainable development in Africa: The difference worth appreciating. In P. B. Cobbinah, & M. Addaney (Eds.), *The geography of climate change adaptation in urban Africa* (pp. 3–26). Palgrave Macmillan.

African Union. (2020). *African strategy on climate change*. African Union.

Averchenkova, A., Fankhauser, S., & Nachmany, M. (Eds.). (2017). *Trends in climate change legislation*. Edward Elgar Publishing.

Bellinson, R., & Chu, E. (2019). Learning pathways and the governance of innovations in urban climate change resilience and adaptation. *Journal of Environmental Policy and Planning, 21*(1), 76–89.

Dow, K., Berkhout, F., Preston, B., et al. (2013). Limits to adaptation. *Nature Climate Change, 3*, 305–307. https://doi.org/10.1038/nclimate1847

Intergovernmental Panel on Climate Change (IPCC). (2021). *Climate change 2021: The physical science basis*. https://www.ipcc.ch/report/ar6/wg1/downloads/report/IPCCAR6WGIFullReportsmaller.pdf. Accessed 21 October 2021.

Intergovernmental Panel on Climate Change (IPCC). (2018). Summary for policymakers. In V. Masson-Delmotte, P. Zhai, H.-O. Pörtner, D. Roberts, J. Skea, P. R. Shukla, A. Pirani, and others (Eds.), *Global warming of 1.5 °C*. World Meteorological Organization.

International Monetary Fund (IMF). (2016). *Small states' resilience to natural disasters and climate change—Role for the IMF* (IMF Policy Paper).

International Monetary Fund. (2021). *Adapting to climate change in Sub-Saharan Africa*. https://www.imf.org/-/media/Files/Publications/REO/AFR/2020/April/English/ch2.ashx

IPCC. (2007). Glossary. In *Climate change: Impacts, adaptation and vulnerability. Contribution of working group II to the fourth assessment report of the Intergovernmental Panel on Climate Change* (pp. 869–883). https://www.ipcc.ch/site/assets/uploads/2018/03/ar4wg2fullreport.pdf

IPCC. (2014). Annex II: Glossary (K. J. Mach, S. Planton, & C. von Stechow, Eds.). In Core Writing Team, R. K. Pachauri, & L. A. Meyer (Eds.), *Climate change 2014: Synthesis report. Contribution of working groups I, II and III to the fifth assessment report of the Intergovernmental Panel on Climate Change* (pp. 117–130). IPCC.

IPCC. (2021). Summary for policymakers. In V. Masson-Delmotte, P. Zhai, A. Pirani, S. L. Connors, C. Péan, S. Berger, N. Caud, Y. Chen, L. Goldfarb, M. I. Gomis, M. Huang, K. Leitzell, E. Lonnoy, J. B. R. Matthews, T. K. Maycock, T. Waterfield, O. Yelekçi, R. Yu, & B. Zhou (Eds.), *Climate change 2021: The physical science basis. Contribution of working group I to the sixth assessment report of the Intergovernmental Panel on Climate Change* (pp. 3–32). Cambridge University Press. https://doi.org/10.1017/9781009157896.001

Marcott, S., Shakun, J., Clark, P., & Mix, A. (2013). A reconstruction of regional and global temperature for the past 11,300 years. *Science, 339*(6124), 1198–1201.

Mayer, B. (2018). *The international law on climate change*. Cambridge University Press.

Nyiwul, L. (2019). Climate change mitigation and adaptation in Africa: Strategies, synergies, and constraints. In T. Sequeira & L. Reis (Eds.), *Climate change and global development*. Springer.

Rigaud, K., de Sherbinin, A., Jones, B., Bergmann, J., Clement, V., Ober, K., Schewe, J., Adamo, S., McCusker, B., Heuser, S., & Midgley, A. (2018). *Groundswell: Preparing for internal climate migration*. World Bank.

Scheffer, M., Carpenter, S., Dakos, V., & van Nes, E. (2015). Generic indicators of ecological resilience: Inferring the chance of a critical transition. *Annual Review of Ecology, Evolution, and Systematics, 46*, 145–167.

Schellnhuber, H. J. (2013). *Turn down the heat: Climate extremes, regional impacts, and the case for resilience*. A Report for the World Bank by the Potsdam Institute for Climate Impact Research and Climate Analytics.

CHAPTER 2

Climate Change Impediments to Realising the Right to Development in Africa

Carol Chi Ngang

1 Introduction

This chapter interrogates the extent to which climate change may pose an impediment to realising the right to development in Africa. The African Charter on Human and Peoples' Rights (1981, art 22) guarantees the right to economic, social and cultural development, achievable with due regard to the equal enjoyment of the common heritage, which is understood to encompass the natural environment and its appurtenance resources that are currently threatened and predisposed to the effects of climate chance. The article 22 provision on the right to development is read in conjunction with article 24, which in actual terms, guarantees

C. C. Ngang (✉)
Department of Public Law, National University of Lesotho, Roma, Lesotho
e-mail: ccngang@nul.ls

Free State Centre for Human Rights, University of the Free State, Bloemfontein, South Africa

M. Addaney et al. (eds.), *Climate Change in Africa*,
https://doi.org/10.1007/978-3-031-30050-9_2

that "[a]ll peoples shall have the right to a general satisfactory environment favourable to their development", importantly because both rights are collective entitlements guaranteed to the peoples of Africa.

Besides the impact of climate change, the rush for and wanton extraction of natural resources in Africa in response to the global competition for economic growth has meant adverse effects on sustainable development on the continent. The cyclical threat of climate change and environmental degradation suggests that prospects for realising the right to development on the basis of a healthy environment is not guaranteed. While it is valid to be apprehensive of this concern, it is noted that Africa's anthropogenic contribution to climate change is less than 5%—the lowest among all regions around the world (Cilliers, 2020, pp. 288–289). Africans on average, contribute a minimal 1.2 tons of carbon emissions per annum as compared to over 16 tons by individual Americans, 12 tons by individual Russians, 9 tons by individual Japanese and 7 tons by individual Chinese (Cilliers, 2020, p. 288). With this reality, the question arises whether Africa should be fretful about climate change, especially considering other defiant challenges that threaten livelihood on the continent.

The obligation to eliminate obstacles to development necessitates closer attention to the climate change impediment particularly because, although Africa's contribution to global warming is minimal, the brunt of it will be borne by Africa more than any other region around the world. Faced with the encroaching threat, Africa has opted for adaptation as a strategy to ensure the continent is acclimatised to the adverse effects of climate change in the immediate, medium and long term (Africa Adaptation Initiative, 2021). Adaptation is conceived here as necessitating adjustments in institutional mechanisms and livelihood patterns as well as putting in place of concrete measures to mitigate the impact of climate change on livelihood. Under the circumstances, this chapter contends that adaptation is insufficient as a strategy for redressing climate change in Africa. It equally necessitates recourse to the law in dealing with contraventions such that environmental justice and fairness is guaranteed to local populations that shoulder the burden of climate change.

The central inquiry revolves around the level of policy responsiveness to Africa's climate change adaptation approach in the midst of competing development challenges. Drawing from the rule of law principle that obligates compliance and adherence to the law (Waldron, 2016), the enquiry is premised on a qualitative socio-legal analysis of relevant legal instruments, case law and related literature. This methodological approach allows for a comprehensive exposition of the obligations that compel African state governments to take appropriate measures to sustain national development endeavours, which entail ensuring that climate change does not disrupt socio-economic and cultural normality on the continent. The analysis and arguments are anchored on the supposition that if environmental laws and climate change conventions are adhered to and appropriate policies and implementation strategies put in place, the threat of climate change would be averted and thus, optimise prospects for achieving the right to development in Africa failing which, recourse should be had to punitive measures.

Contemporary approaches to climate change vary considerably, with contentious debates on what approach is most appropriate. Some proponents advocate for the principle of common but differentiate responsibility and respective capabilities (Pauw et al., 2014) and the polluter pays principle (Cordato, 2001), which developed countries led by the US are vehemently opposed to, in favour of the less stringent no-harm rule (Birnie et al., 2009, pp. 143–152; Brownlie, 2008, pp. 275–285) and the principle of prevention (Duvic-Paoli, 2018, pp. 1–12), which unfortunately, have not had much positive impact. With respect to Africa's adaptation approach, which as indicated earlier is inadequate in responding to the effects of climate change, this chapter posits that a rigorous punitive sanctions regime targeting developed countries whose uncontrollable industrialisation and resultant carbon emissions are known to cause the increasing wave of global warming in contravention of applicable conventions, the consequences of which impede realisation of the right to development in Africa.

Despite the commitments undertaken under various framework mechanisms and binding treaty instruments, such as the Paris Agreement to limit global warming to below 2 degree Celsius, compared to pre-industrial levels, the reality speaks to the contrary, raising peripheral questions relating to envisaged sustainable development targets for the African continent. The setbacks to development resulting from the climate change emergency discussed in Sect. 2 necessitates a closer examination of the context for sustainable development in Africa, which entails a commitment to the right to development and the right to a satisfactory environment as illustrated in Sect. 3. Based on the findings, the policy recommendations preferred in the concluding section, build on the legal prescriptions that enjoin African state governments to take actionable measures, including the crafting of a sanctions regime against developed countries that perpetuate and sustain climate change.

2 Climate Change Emergency

Climate science is unambiguous about the fact that the world is entrapped in a climate emergency necessitating urgent action (Ripple et al., 2021, pp. 894–898). Climate change does not only pose an emergency hazard; the many uncertainties in predictions of it with regard to timing, magnitude and regional patterns thereof, make it even more worrying for Africa where, owing to extreme poverty and minimal capacity for adaptation, the converging pressures on livelihood compromises prospects for sustainable development.

2.1 Converging Pressures on Livelihood

Climate change is triggered by two major factors; the biogeochemical (natural events and processes) and anthropogenic influences (human activities) (Odjugo, 2009, p. 61). A detailed discussion of these factors is outside the scope of this chapter. It is however, commonly agreed that the anthropogenic factors result from global warming that occasions climate change (Wang & Chameides, 2005, p. 39). The World Health Organisation (WHO) (2014) estimates that between 2030 and 2050, climate change would cause about 250,000 deaths annually. The Food and Agricultural Organisation (2008, p. 8) equally warns that agricultural productivity will plummet drastically as a result of climate change and thus, affect food security in large parts of Africa. While industrialised countries that generate much of the carbon emissions that cause global warming possess the capacity and the coping mechanisms to deal with the effects of climate change (Mcguian et al., 2002), developing regions of the world like Africa that shoulder the weight of the problem, generally do not, which is a major concern.

Geographically located in the tropics with low-lying topography, Africa stands at the frontline, exposed to freak weather conditions that are projected to cause massive displacement of an estimated 150 million people across the continent by 2050 (Addaney et al., 2017, p. 6). It is noted that climate change has since the 1970s, caused intense rainstorms accompanied by widespread flooding and extremely severe droughts across Africa (Africa Centre for Strategic Studies, 2021). It means that without climate change, these events would either not have occurred or that the incidence would have been less frequent and the magnitude and intensity of them less devastating. The map (Fig. 1) provides a geographical representation of the extent to which droughts and floods resulting from climate change impact on livelihood in Africa.

Northern Kenya is currently affected by one of the most protracted and intense droughts that the country has experience in the last couple of decades. Dependent largely on rainfall for cattle breeding, food crop cultivation and for socio-economic development, disruptions in the rainfall pattern since December 2020 have pushed the impoverished pastoralist communities and the livestock they depend on for survival to the brink of disaster (Pietromarchi, 2021). Livelihood for an estimated 2.4 million

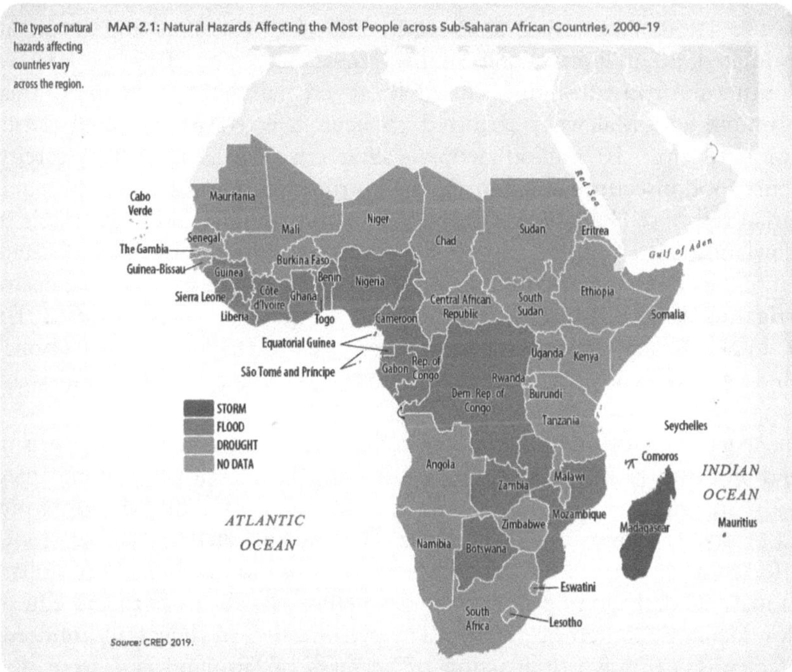

Fig. 1 The effects of climate change in Africa (*Source* World Bank Group—Africa Pulse, October 2021 Edition)

people, including 465,000 children and 93,000 pregnant women and breastfeeding mothers are threatened by escalating poverty, malnourishment, water scarcity, food insecurity and starvation (Pietromarchi, 2021). Although Kenya is naturally prone to recurrent droughts with 80% of its territory consisting of arid or semi-arid land with only sporadic rainfalls, Kandji (2006, pp. 17–18) notes that global warming has increasingly aggravated the situation, making it a hurdle to achieving the country's sustainable development goals. Close to 10 million people of the Kenyan population live in the drought-prone regions and consequently, competition for scarce resources exacerbates environmental degradation, which in turn increases their vulnerability to drought (Kandji, 2006, p. 19).

Beyond Kenya, worsening droughts have also left devastating effects in the Sahel region that stretches across Sudan, Chad, Niger, Burkina Faso, Mali, Mauritania and Senegal. In the Lake Chad basin, the shrinking lake (which used to be the largest freshwater body on the continent in the 1960s), is having a direct livelihood-threatening impact on an estimated 50 million people in the region (Bishop, 2017, p. 83). In Southern Africa, droughts caused by severe rainfall deficits in Angola, Namibia and Malawi is reported to have affected a combined population of over 10 million people facing crisis levels of water scarcity, acute food insecurity and chronic malnutrition, requiring urgent humanitarian response (Reliefweb, 2021). While humanitarian aid is inevitable in alleviating suffering, the targets for sustainable development, necessitate more development-oriented responses in terms of, for example, extensive irrigation of the drought-prone regions. Otherwise, as Tadesse (2010, p. 1) points out, climate change will continue to challenge Africa's policy choices and ability to achieve sets targets and goals under framework programmes for development.

Apart from droughts, Africa is confronted with recurrent rainstorms and floods. In 2019, two consecutive tropical cyclones accompanied by hard winds, heavy rainstorms and raging floods hit Mozambique, Comoros, Malawi and Zimbabwe (Reliefweb, 2019). The cyclones affected an estimated 3.8 million people in addition to billions of dollars of material damage to property, infrastructure as well as health and education facilities (Reliefweb, 2018). The cyclones left in their trail hundreds of deaths, extensive destruction of hectares of farmland and crops and massive displacement of the local populations leaving hundreds of thousands, including children in distress and in dire need of humanitarian assistance (United Nations Children's Fund [UNICEF], 2019; World

Vision, 2019). Mozambique alone, hardest hit by the 2019 cyclone saw some 3,000 km^2 of its land surface submerged in floods with close to 285,000 homes completely destroyed (Reliefweb, 2018).

The cyclones, which are reported to be fed by warming of ocean temperatures, rising sea levels and extreme weather conditions have been described as the worst natural disaster to hit Southern Africa in the last two decades (UNICEF, 2019). It is worth noting that Comoros and Zimbabwe rank as lower-medium-income countries while Mozambique and Malawi rank in the low-income grouping, placing them among the extremely poor and least developed countries in Africa and the world. The poverty levels in these countries obliterate their resilience to climate change and the capacity to respond adequately to the cyclones, which are projected to become increasingly frequent as global warming intensifies. Before the affected regions could fully recover from the shockwaves of the 2019 cyclones, another cyclone made landfall in Mozambique in January 2021 with Madagascar and parts of Zimbabwe, South Africa and Botswana also severely affected (World Meteorological Organisation, 2021). With the frequency and magnitude of climate events, it is hard to imagine that the right to development would be achieved in Africa.

2.2 *The Right to Development in Africa*

The African Charter (1981, art 22) enshrines the provision that all the peoples of Africa are entitled to better living standards resulting from the processes of socio-economic and cultural development. The Declaration on the Right to Development (DRTD) (1986, art 1(1)) defines development as an inalienable human right that entitles every human person and all peoples to participate in, contribute to and share the benefits that obtain from the process, as a guarantee for the full realisation of all other human rights and fundamental freedoms. For development to be holistic and equitably beneficial, requires a natural environment that is generally satisfactory (African Charter, 1981, art 24). The duty to individually or collectively facilitate the processes for development to the benefit of their peoples compels African state governments to adopt appropriate policies and eliminate obstacles such as those caused by climate change that compromise well-being. When the effects of climate change present as illustrated in the previous section, the brunt is shouldered principally by the poverty-stricken peoples who make up a huge proportion of the African population.

Contrary to the emphasis on the social, economic and environmental dimensions to development under the Sustainable Development Goals (SDG) (UN General Assembly, 2015), the frequency of environmental hazards resulting from climate change suggests that the right to development in Africa may not be attained. Realisation of the right to development, which should translate into improved well-being for the peoples of Africa, ought to be progressive and incremental and thus, disallows regression in the enjoyment of existing entitlements (Ngang, 2021a, pp. 282–287). Climate events such as those described in the previous section not only cause regression in living standards but also threatens the existence of vulnerable populations that remain exposed to environmental calamities. Even as African state governments may not be in a position to halt climate change, they do have the legal capacity and indeed, are obligated by law to take actionable measures against perpetrators considering that the human-induced factors that trigger climate change originate from contravention of the laws that proscribe wrongful actions and practices that generate and sustain climate change. Wewerinke-Singh (2019) notes that climate change has increasingly impacted on the enjoyment of a broad range of human rights particularly owing to lack of a comprehensive understanding of the role of international human rights law in invoking state responsibility and accountability for actions related to climate change.

2.3 *The Rule of Law Imperative*

The rule of law is both a legal and a governance principle in respect whereof states and non-state actors are equally subjected to legal accountability under clearly formulated legal systems and processes. It literally infers an understanding of the law as the governing instrument that binds every functionary of society. The rule of law entails adherence to the law and therefore, forbids any infringement of it. It is intended to ensure that justice prevails and consequently, allows recourse to legal checks and balances. In essence, the rule of law safeguards against impunity to the effect that when the law is contravened, legal action can be undertaken as it happened in the environmentally related charges against Kenya in the *Ogiek Community* (2017) and *Endorois* (2009) cases and against Nigeria in the *SERAC* (2001) case. It imposes obligations on states as custodians of the law, to institutionalise legislative, policy and regulatory measures as a caution that to act outside of the legal prescriptions would amount to undermining the law.

Applied in context to the processes for development and environmental protection, the African Charter (1981, art 1), in addition to enshrining the right to development and the right to a general satisfactory environment favourable for development, enjoins state parties to adopt legislative or other measures to give effect to the broad range of human and peoples' rights enshrined in the Charter. The obligation extends to regulating non-state actors whose operations often have environmental consequences on local populations. Some of these operations have ended up in litigation albeit in foreign courts against Shell Petroleum Development Company (SPDC) Ltd. on the basis of its activities in Nigeria (*Okpabi v Royal Dutch Shell*, 2021) and Vedanta Resources PLC operating in Zambia (*Lungowe v Vedanta Resources*, 2017). More on these cases are discussed below to support the fact that the rule of law entitles affected communities to hold non-state entities to legal accountability for the effect of their actions on the environment and on livelihood.

The obligation to take legislative and other measures includes not only those at the domestic level but also undertakings under international law such as the DRTD adopted in 1986 and the Resolution on the Right to a Clean, Healthy and Sustainable Environment adopted in 2021. Although UN declarations and resolutions are by nature not legally binding, they set the pace for general principles of law that provide the basis for customary international law, which by virtues of article 38 of the Statute of the International Court of Justice, is a recognised source of public international law. With regard to the right to development, Bunn (2000, p. 1460) opines that it embodies the rule of law, which constitutes a major guiding principle for its realisation and thus, to act outside of it would amount to undermining law. The Office of the High Commissioner for Human Rights (2018) observes that the right to a healthy environment provides the framework for more effective enforcement of environmental laws and policies, greater public participation in environmental decision-making, reduced environmental injustices and a level playing field for the realisation of socio-economic and cultural rights.

Although not stated explicitly, the right to a satisfactory environment is contained in a broad range of international law instruments on climate change, biodiversity conservation, land management, marine environment, ozone layer protection, chemical and waste management among others (United Nations Environment Programme, 2007, pp. 34–100). It has become an accepted practice within the framework of environmental law that polluters of the environment must shoulder the remediation and compensation costs of the resultant impact on livelihood (European Court of Auditors, 2021, pp. 6–7; Moffet & Bregha, 1996, p. 6). Considering Africa's relatively minimal contribution to climate change, the polluter pays principle guarantees that the loss and damage incurred on the continent is charged to developed countries for their excessive carbon emissions. In 2015, world leaders reiterated the commitment to "take urgent action to combat climate change and its impacts" (UN General Assembly, 2015, goal 13). To this end, the SDG sets an achievable target to:

> Implement the commitment undertaken by developed country parties to the United Nations Framework Convention on Climate Change to a goal of mobilizing jointly $100 billion annually by 2020 from all sources to address the needs of developing countries in the context of meaningful mitigation actions and transparency on implementation and fully operationalise the Green Climate Fund through its capitalization as soon as possible. (UN General Assembly, 2015, target 13a)

Even though the SDG instrument is by nature not legally binding, its implementation is envisaged within the framework of international law (UN General Assembly, 2015, para 18), which ought to be read in conjunction with the legally binding UN Framework Convention on Climate Change [UNFCCC] (1992) and the Paris Agreement (2015) that regulate climate action globally. The UNFCCC (1992, arts 3/1&(2)) obligates "developed country Parties [to] take the lead in combating climate change and the adverse effects thereof" with full consideration given to the "specific needs and special circumstances of developing country Parties, especially those that are particularly vulnerable to the adverse effects of climate change, and of those Parties, especially developing country Parties, that would have to bear a disproportionate or abnormal burden under the Convention". Climate justice in this instance obliges developed countries to pay for the loss and damage caused by climate change as envisaged in the Warsaw Loss and Damage Mechanism (2013), which unfortunately, is just a soft tool with no binding effect. Ohdedar (2016, pp. 20–40) notes that without clearly defined obligations and beneficiary rights, it will be difficult to hold defaulters to account under the Warsaw Mechanism.

The Paris Agreement enshrines the commitment to generate an annual base level funding of USD100 billion aimed at off-setting the adverse effects of climate change with particular focus on the needs and priorities of developing countries (Conference of the Parties, 2015, para 5). This commitment as stated in the Kyoto Protocol principally "only binds developed countries, and places a heavier burden on them under the principle of 'common but differentiated responsibility and respective capabilities', because it recognizes that they are largely responsible for the current high levels of GHG emissions in the atmosphere" (UN Climate Change, 2021). The framing of these actions as legal commitments presents, according to the Global Development Policy Centre (2021), an opportunity to build a development-centred productive growth and stability model that allows for adaptation to the changing climate while equalising opportunities for development and raising levels of prosperity across the world.

Without overlooking other approaches in combating climate change, financing in the form of compensation for the resultant loss and damage is of foremost importance to Africa as a resilience building measure. The Glasgow Climate Pact (2021, para 30) "*[e]mphasizes* the challenges faced by many developing country Parties in accessing finance and *encourages* further efforts to enhance access to finance, including by the operating entities of the Financial Mechanisms". It is observed as part of the negotiations at the COP26 summit that "[d]eveloped countries have made progress towards delivering the $100 billion climate finance goal and will reach it by 2023 at the latest" (UN Climate Change Conference, 2021). The promised funding is envisaged to enable vulnerable developing countries beef-up their adaptive capacities, reduce exposure to the adversities of climate change and thus, creates legitimate expectations of positive concrete action in sustaining development in Africa.

The implication of having commitments on climate action enshrined in legal instruments is to the effect that when a violation is established, not excluding default in financial obligations, recourse to remedial actions can be pursued. Unfortunately, despite the range of international legal instruments that are designed and intended to regulate environmentally hazardous activities and actions that perpetuate global warming and consequently climate change, there are as pointed out earlier, no remedial mechanisms at the international level through which redress could be sought in the event of a contravention. With the prevailing trend, the tipping point may eventually see whole communities in Africa wiped out of existence by drought and flooding among other global warming-induced catastrophes, raising questions with regard to the targets for sustainable development in Africa.

3 The Context for Sustainable Development in Africa

Sustainable development, which Moellendorf (2011, pp. 445–450) perceives as a claimable human right, entails as Gupta and Vegelin (2016, p. 438) observe, an emphasis on environmental inclusiveness; conceived in terms of the bond between impoverished communities and the local ecosystem on which they depend for survival and accordingly, how their vulnerability is exacerbated by the effects of climate change. To this end, sustainable development in Africa entails combining the mutually reinforcing right to development and the right to a general satisfactory environment enshrined in the African Charter. The right to development guarantees that in the absence of any form of impediment, its realisation will lead to better living standards and improved well-being at the economic, social and cultural levels. The environmental dimension requires that the processes for development be sustained through a governance dispensation that is sufficiently enabling.

3.1 Enabling Governance Dispensation

The context for sustainable development, with emphasis on environmental inclusiveness as noted above, invokes the right to development and the right to an environment that is conducive for development, which obligates African state governments to create an enabling governance dispensation wherein the processes for development do not impact adversely on the environment and that the environment does not impede prospects for development. It sums up to the principle of intergenerational equity that needs to be sustained by a governance policy framework that guarantees equitable management of available resources to the benefit of present and future generations (Ebuke, 2003, p. 2; Ngang, 2021b, pp. 95–97). Formulated as legally binding prescriptions that underpin the governance policy framework for sustainable development, the right to development and the right to a satisfactory environment enjoin state governments to prioritise the common good and the socio-economic and cultural development exigencies of the peoples of Africa and hence, the obligation to eliminate impediments to the development process (DRTD, 1986, art 6(3)).

Although the Declaration on the Right to Development does not mention climate change as an impediment, a comprehensive reading of the concept of the right to development, which is concerned about the constant improvement in human well-being, entails that conceivable obstacles that may compromise the processes for development be eradicated. As already highlighted, realisation of the right to development is contingent in part on a general satisfactory environment, referring to one that is free of hurdles such as those resulting from climate change. Climate change disrupts socio-economic and cultural development and the livelihood entitlements associated thereto. It wipes out existing development gains and also hinders prospects for further development. The Malé Declaration (2007) affirms that climate change has immediate implications for the full enjoyment of a range of human rights. The African Charter accordingly, enjoins state parties to take standard-setting measures in ensuring that the right to development is achieved and that the environment is equally suitably conducive for development to take place without impediments.

Article 1 of the African Charter compels state parties to "adopt legislative or other measures to give effect to [...]" the rights, duties and freedoms enshrined therein, including in relation to article 22, the "[...] duty, individually or collectively, to ensure the exercise of the right to development" and the duty associated with article 24, to give effect to "the right to a general satisfactory environment favourable for [...] development". All 55 African states have ratified the Charter, which according to Ouguergouz (2003, p. 37), set standards against violations and therefore, obligates state parties to put in place a regulatory framework that instructs against regression in the enjoyment of the rights and freedom contained therein. The DRTD (1986, art 2(3)) ascribes to states, the right and the duty to formulate suitable development policies to ensure constant improvement in well-being. The African Union (AU) Constitutive Act (2000, art3(d)) enjoins member states to "[p]romote and defend African common positions on issues of interest to the continent and its peoples". Climate change is one of those major concerns for which AU member states are obligated to adopt a common policy position, including with regard to crafting a sanctions regime as illustrated below.

The context envisages a governance dispensation that combines political and legal actions. So far, the AU has in terms of political action, put in place a comprehensive Climate Change Strategy (adopted in 2014 and revised in 2020) within the broader framework of Agenda 2063, which assures of adequate response, coping capacity and resilience to the adversities of climate change (Africa Union, 2020). While the climate change strategy is a sensible move in the right direction, unless actionable measures are taken, the growing patterns of intense droughts and ravaging floods and the loss and damage they cause are likely to continue to spiral into calamities that impede actualisation of the right to development for the present and future generations in Africa. The context for sustainable development thus, necessitates recourse to the prescriptions of the law in terms of which the processes for holding climate change perpetrators lawfully accountable could be pursued.

3.2 Climate Change Litigation

As earlier stated, the duty is imputed on African state governments to regulate the operations of non-state entities that may contravene provisions of the African Charter. Otherwise, legal accountability reverts to the states in the event that they fail to do so. In the *SERAC* (2001, paras 50–54) litigation, the Nigerian government was held accountable before the African Commission for environmental offences committed in Ogoniland by Shell Corporation whose oil exploration activities the government complicitly failed to regulate. In finding the Nigerian government in violation of the right to a general satisfactory environment, the African Commission highlighted that state governments are under an obligation to protect their citizens from pollution and environmental degradation (*SERAC*, 2001, para 52). Other African state governments have equally not demonstrated sufficient commitment in regulating foreign multinationals or in holding them accountable for the environmental offences they commit across the continent.

In the absence of domestic remedy mechanisms, affected local communities have demonstrated that claims could be pursued in foreign courts against parent companies to those operating on the continent. In the *Okpabi v Royal Dutch Shell* case, 40,000 Niger Delta community members brought action before the UK Supreme Court against Royal Dutch Shell for the environmental damage and human rights abuses committed by its Nigerian subsidiary, SPDC Ltd. Admissibility of the case made Royal Dutch Shell answerable for the offences committed by its Nigerian subsidiary (Burianski et al., 2021, p. 4). In *Lungowe v Vedanta Resources*, the UK Supreme Court also declared admissible a claim brought by a group of local farmers from Zambia against the UK-based Vedanta Resources PLC, alleging extensive loss and damage suffered as a result of environmental degradation caused by the abusive operation of its Zambian subsidiary, *Konkola Copper Mines Plc*. The merits of both cases are yet to be determined. In the backdrop of a rise in climate change litigation globally, in spite of Africa's disproportionate exposure to the hazards, only a few climate change cases have been recorded on the continent, probably owing to functionally weak or non-existent legislative framework and the practical constraints that claimants face in seeking access to justice (Burianski et al., 2021, pp. 4–6).

Many African countries either do not have robust domestic legislation for regulating environment crimes or those that are available are practically non-functional in guaranteeing effective protection to victims of environmental offences. In Nigeria, even with the National Environmental Standards Regulations and Enforcement Agency (Establishment) Act of 2007 and related regulations (Sodipo et al., 2017), the *Okpabi community could not find domestic remedy to their ordeals and had to resort to a foreign jurisdiction for redress.* In the Zambian case, it cannot be argued otherwise. If the Zambia legal system was protective enough, the claimants in the *Vedanta Resources* case would have been capable of obtaining justice from domestic. However, in upholding the claimants' *locus standi* to seek remedy in the UK, the UK Supreme Court noted that "the claimants would not obtain substantial justice if required to proceed [with their claim] in Zambia" (*Lungowe v Vedanta Resources*, 2019, para 84). While the legal systems in Africa do not provide effective domestic remedy against multinationals for the environmental crimes they commit in Africa, the tough challenge may be even more with regard to taking legal actions against developed countries when they contravene their obligations under the various climate change conventions and agreements.

3.3 Considering Climate Sanctions

The rule of law obligates states to honour their commitments on climate action. When the rules are contravened, parties are justified to have recourse to other remedy mechanisms (African Development Bank, 2011), including the use of sanctions. The European Union (EU) has for instance, warned that "as of 2024, it will have the power to withdraw preferential trade access for developing countries that do not meet the standards for climate action set by [the] Paris and other environmental agreements" (World Economic Forum, 2021). It gives the wrong impression that only developing countries are defaulters of climate action. Against the fact that developed countries contribute heavily to climate change and are unlikely to cut back on their carbon emissions, it is patronising that the EU finds justification to impose sanctions on developing countries but does not consider doing so to its member states and other developed countries that default in honouring their commitments.

Sanctions are conceived under public international law as designating "a broad range of reactions adopted unilaterally or collectively by the States against the perpetrator of an internationally unlawful act in order to ensure respect for and performance of a right or obligation" (Decaux, 2008, p. 249). With the existing geopolitical configuration that defines relations between states wherein, Africa remains intimately dependent on and heavily indebted to developed countries, the likelihood is lean that Africa could effectively impose sanctions that have any bearing on developed countries. Doing so may attract retaliatory sanctions with uncomfortable boomerang effects. However, when crafted and applied diligently, sanctions can be a helpful legal instrument, with growing inclination towards redressing human rights violations. Slack (2021) affirms that sanctions can play a strategic role in averting climate change, which Addaney et al. (2017, pp. 12–27) argue, is a human rights problem. Perceived as such, sanctions are legitimate and hence, not limited to be used exclusively by developed countries. Africa is equally equipped with the legitimacy to craft a sanctions regime against developed countries that perpetuate and sustain climate change.

It is acknowledged that climate change cannot directly be linked to any particular country by virtue of the fact that it is generated from multiple sources. A strong argument has however, been advanced to the effect that vulnerable countries have a *prima facie* case to bring legal claims against developed countries for damages caused by climate change as a result of the metric tons of carbon they emit into the atmosphere (Ohdedar, 2016, p. 20; Voigt, 2008, p. 1). To hold these countries to account, Hagen and Schneider (2021, pp. 1–3) suggest the use of trade sanctions, including the imposition of import tariffs as a punitive sanctions mechanism against defaulting countries. This could be achieved through a number of possible approaches: First, counting on the fact that Africa is home to a myriad of natural resources that are of enormous economic value to industrialisation in developed countries (African Development Bank, 2011, p. 25), selective trade embargoes could be imposed on some of the strategic resources that developed countries desperately need to sustain their economies. Second, Africa could impose punitive import tariffs or prohibit buying certain commodities that should have been tagged with a "climate change blacklisting", especially if substitute commodities could be sourced from other countries that are climate compliant. A third option would be to revisit some of the trade agreements between AU member states and climate defaulting developed countries and probably retract or revise the agreements. Cirone and Urpelainen (2013, p. 310) affirm that trade sanctions have the potential to reduce the profitability of "free riding" by climate defaulters and thus, improve compliance with environmental treaties and climate change conventions.

Cognisance is taken of the fact that retaliatory sanctions could have a harsher effect, which to guard against, requires of Africa to work on building self-sufficiency in major sectors of the economy such as currency sovereignty, food production and processing as well as strategic supplies in the area of energy among others. The enforcement of trade sanctions as Hagen and Schneider (2021, p. 2) opine, requires a stable coalition, which in this instance is attainable within the framework of the AU and the African Continental Free Trade Area (AfCFTA) which have the capacity to ensure policy stability and collective enforcement. As an emerging economic powerhouse, Africa is steadily gaining the leverage to determine terms of trade and thus, shape global policy-making and hence, the opportunity to maximise resilience against climate change and the capacity to endure contempt and bullying by developed countries. Africa would equally need to take advantage of the growing climate change activism by civil society movements on the continent and globally, to be able to exert the requisite pressure on defaulting developed countries. It might be expedient as Mulder (2019) argues, to direct climate sanctions at companies that actually emit greenhouse gases rather than target countries. However, with respect to the law of treaties, states have the primary responsibility to abide by the climate change conventions they have committed to under international law, which obligates them to regulate the operations of their home companies.

4 Conclusion

The loss and damage that continues to pill up against impoverished communities and vulnerable populations across Africa as a result of climate change as illustrated in this chapter, does not only breach the right to a generally satisfactory environment but indeed, also impede prospects for realising the right to development. As illustrated, the right to development holds the promise that its realisation would create the opportunity for better standards of living and improved well-being for the peoples of Africa, which, however, stands threatened by the adverse effects of climate change. The findings in this regard draw essentially from a comprehensive reading of provisions of the legal instruments that regulate climate action, which necessitates applying the rule of law in compelling defaulting developing countries to honour their undertakings on climate action under applicable conventions. Although Africa's contribution to climate change is comparatively negligible, the continent has suffered and remains susceptible to greater distress even as state governments struggle to make do with resilience and adaptation models in responding to climate change.

In terms of the UNFCCC and its operational protocols and agreements, financing to defray costs of the devastating effects of climate change, calculated at an annual base limit of $100 billion are charged to developed countries. These amounts are due and payable to developing countries and accordingly, create legitimate expectations for Africa, which has suffered and remained susceptible to climate change-induced loss and damage. Developed countries have undertaken to make the promised financing available to developing countries as funding for their adaptation strategies. When they default in this commitment, they breach international rules and therefore, become predisposed to legal accountability, including through climate change litigation which is yet to be explored to full extent in Africa. In the absence of any international remedy mechanism through which redress could be sought, the argument is advanced in this chapter that recourse could be had to alternative measures like selective trade sanctions on export of strategic natural resources and/or import tariffs on goods from developed countries that perpetuate climate change. The rule of law obligates adherence to the obligations imposed by climate change conventions. As an aggrieved party, Africa is equipped with the legitimacy and has the capacity within the framework of the AfCFTA to craft enforceable trade sanctions against developed countries that fail to honour their commitments on climate action. The challenge in doing so is real but not unattainable.

References

Addaney, M., Boshoff, E., & Olutola, B. (2017). The climate change and the human rights nexus in Africa. *Amsterdam Law Forum, 9*(3), 5–28.

Africa Adaptation Initiative. (2021, November). https://africaadaptationinitiative.org/

Africa Centre for Strategic Studies. (2021, October). *How global warming threatens human security in Africa*. https://africacenter.org/spotlight/how-global-warming-threatens-human-security-in-africa/#:~:text=According%20to%20the%20Intergovernmental%20Panel,the%20Sahel%20and%20southern%20Africa

African Charter on Human and Peoples' Rights adopted in Nairobi, Kenya on 27 June 1981.OAU Doc CAB/LEG/67/3 Rev.5(1981).

African Commission on Human and Peoples' Rights (Ogiek Community) v Republic of Kenya (2017) Appl No 006/2017.

African Development Bank. (2011). *Africa in 50 years time: The road towards inclusive growth*. African Development Bank Group.

African Union. (2020). *African climate change strategy 2020–2030* (Revised Draft).

Birnie, P., Boyle, A., & Redgwell, C. (2009). *International law and the environment* (3rd ed.). Oxford University Press.

Bishop, R. (2017). African leadership in a time of climate risk. In *Confronting climate change: Africa's leadership on an increasingly urgent issue* (Issue Brief). Overseas Development Institute.

Brownlie, I. (2008). *Principles of public international law* (7th ed.). Oxford University Press.

Bunn, I. D. (2000). The right to development: Implications for international economic law. *American University of International Law Review, 15*(6), 1425–1467.

Burianski, M., Clarke, M., Kuhnle, F. P., & Wackwitz, G. (2021). Climate change litigation in Africa: Current status and future developments. *Africa Focus, 7*, 2–9.

Centre for Minority Rights Development (Kenya) & Minority Rights Group International on behalf of Endorois Welfare Council v Kenya Comm 276/2003 (2009) AHRLR 75 (ACHPR 2009).

Cilliers, J. (2020). *Africa first: Igniting a growth revolution*. Jonathan Ball Publishers.

Cirone, A. E., & Urpelainen, J. (2013). Trade sanctions in international environmental policy: Deterring or encouraging free riding? *Conflict Management and Peace Science, 30*(4), 309–334.

Conference of the Parties 'Adoption of the Paris Agreement' Twenty-first session Paris, 30 November to 11 December 2015, United Nations FCCC/CP/2015/L.9/Rev.1.

Constitutive Act of the African Union adopted in Lomé, Togo on 11 July 2000.
Cordato, R. E. (2001). *The polluter pays principle: A proper guide for environmental policy* (No. 6). Institute for Research on the Economics of Taxation—Studies in Social Cost, Regulation, and the Environment.
Decaux, E. (2008). The definition of traditional sanctions: Their scope and characteristics. *International Review of the Red Cross, 90*(870), 249–257.
Declaration on the Right to Development Resolution A/RES/41/128 adopted by the UNGA on 4 December 1986.
Duvic-Paoli, L.-A. (2018). Introduction. In L.-A. Duvic-Paoli (Ed.), *The prevention principle in international environmental law* (pp. 1–12). Cambridge University Press.
Ebuke, K. S. A. (2003). The right to a satisfactory environment and the African Commission. *African Human Rights Law Journal, 3*(1), 149–166.
European Court of Auditors. (2021). *The polluter pays principle: Inconsistent application a cross EU environmental policies and actions* (Special Report No 12). Publications Office of the European Union.
Food and Agriculture Organisation. (2008). *Climate change and food security: A framework document*. Food and Agriculture Organisation.
Glasgow Climate Pact, Decision -/CP.26 (advanced unedited volume) adopted at the 26th Summit of the UN Climate Conference (COP26) in Glasgow, United Kingdom 2021.
Global Development Policy Centre. (2021, September). *Toward development-centered climate change policy at the International Monetary Fund*. https://www.bu.edu/gdp/2021/09/23/toward-development-centered-climate-policy-at-the-international-monetary-fund/
Gupta, J., & Vegelin, C. (2016). Sustainable development goals and inclusive development. *Politics, Law and Economics, 16*(1), 433–448.
Hagen, A., & Schneider, J. (2021). Trade sanctions and the stability of climate coalitions. *Journal of Environmental Economics and Management, 109*(102504), 1–16.
Kandji, S. T. (2006). *Drought in Kenya: Climatic, economic and socio-political factors*. Today – New Standpoints.
Kyoto Protocol to the United Nations Convention on Climate Change FCCC/CP/1997/L.7/Add.1 adopted on 10 December 1997.
Lungowe & Others v Vedanta Resources PLC and Another [2019] UKSC 2 UKSC 2017/0185.
Malé Declaration on the Human Dimension of Global Climate Change adopted by Small Island Development Countries at Male', Republic of Maldives on 14 November 2007.

Mcguian, C., Reynolds, R., & Wiedmer, D. (2002). *Poverty and climate change: Assessing impacts in developing countries and the initiatives of the international community*. London School of Economics Consultancy Project for the Overseas Development Institute.

Moellendorf, D. (2011). A right to sustainable development. *The Monist—Morality and Climate Change, 94*(3), 433–452.

Moffet, J., & Bregha, F. (1996). The role of law in the promotion of sustainable development. *Journal of Environmental Law and Practice, 6*(1), 1–22.

Mulder, N. (2019). *Can 'climate sanctions' save the planet?: Economic pressure can be a force for environmental good—If it targets companies, not countries*. https://www.thenation.com/article/archive/climate-green-new-deal/

Ngang, C. C. (2021a). *The right to development in Africa*. Brill.

Ngang, C. C. (2021b). Urban landownership and the right to sustainable development for women in Africa. In R. Home (Ed.), *Land issues for urban governance in Sub-Saharan Africa* (pp. 95–97). Springer Nature Switzerland AG.

Odjugo, P. A. O. (2009). The impact of climate change on water resources: Global and Nigerian analysis. *FUTY Journal of the Environment, 4*(1), 59–77.

Ohdedar, B. (2016). Loss and damage from the impacts of climate change: A framework for implementation. *Nordic Journal of International Law, 85*(1), 1–40.

Okpabi & Others v Royal Dutch Shell & Another [2021] UKSC 3.

Ouguergouz, F. (2003). *African charter on human and peoples' rights: A comprehensive agenda for human dignity and sustainable democracy in Africa*. Kluwer Law International.

Pauw, P., et al. (2014). *Different perspectives on differentiated responsibilities* (Discussion Paper 6/2014). German Development Institute.

Pietromarchi, V. (2021, November). *'We will all die': In Kenya, prolonged drought takes heavy toll*. https://www.aljazeera.com/news/2021/11/17/we-will-all-die-in-kenya-prolonged-drought-takes-heavy-toll

Reliefweb. (2018). *Southern Africa: Drought—2018–2021*. https://reliefweb.int/disaster/dr-2018-000429-zwe

Reliefweb. (2019). *Southern Africa: Cyclones Idai and Kenneth snapshot, as of 10 July 2019*. https://reliefweb.int/report/mozambique/southern-africa-cyclones-idai-and-kenneth-snapshot-10-july-2019

Reliefweb. (2021). *Southern Africa: Drought—2018–2022*. https://reliefweb.int/disaster/dr-2018-000429-zwe#countries

Ripple, W. J., et al. (2021). World scientists' warning of a climate emergency 2021. *BioScience, 71*(9), 894–898.

Right to a Clean, Healthy and Sustainable Environment Resolution 48/13 adopted by the UN Human Rights Council on 8 October 2021.

Slack, K. (2021). *It's time to use sanctions to stop climate destruction*. Earth Rights International. https://earthrights.org/blog/its-time-to-use-sanctions-to-stop-climate-destruction/

Social and Economic Rights Action (SERAC) and Centre for Economic and Social Rights Centre (CESR) v Nigeria Comm 155/96 (2001).

Sodipo, E., et al. (2017). *Environmental law and practice in Nigeria: Overview*. https://uk.practicallaw.thomsonreuters.com/w-006-3572?transitionType=Default&contextData=(sc.Default)&firstPage=true#:~:text=National%20Environmental%20Standards%20Regulations%20and,20)%20and%20repealed%20the%20Federal

Tadesse, D. (2010). *The impact of climate change in Africa* (Institute for Security Studies Paper 220).

United Nation General Assembly. (2015, September). *Transforming our world: The 2030 agenda for sustainable development resolution A/RES/70/1 adopted by the General Assembly*.

United Nations Children's Fund. (2019). *Massive flooding in Mozambique, Malawi and Zimbabwe*. https://www.unicef.org/stories/massive-flooding-malawi-mozambique-and-zimbabwe

United Nations Climate Change Conference. (2021, November). *COP26—Glasgow Climate Pact*. Glasgow, United Kingdom.

United Nations Climate Change. (2021). *What is the Kyoto Protocol?* https://unfccc.int/kyoto_protocol.

United Nations Environment Programme. (2007). *Compliance mechanism under selected multilateral environmental agreements*. UNEP Division of Environmental Law and Conventions.

United Nations Environment Programme. *Facts about the climate emergency*. https://www.unep.org/explore-topics/climate-action/facts-about-climate-emergency.

United Nations Framework Convention on Climate Change adopted by the General Assembly on 12 June 1992.

United Nations Office of the High Commissioner for Human Rights. (2018). *Right to a healthy and sustainable environment report*. https://www.ohchr.org/EN/Issues/Environment/SREnvironment/Pages/HealthySustainable.aspx

Voigt, C. (2008). State responsibility for climate change damages. *Nordic Journal of International Law, 77*(1–2), 1–22.

Waldron, J. (2016). *The rule of law*. Stanford Encyclopedia of Philosophy. https://plato.stanford.edu/entries/rule-of-law/

Wang, J., & Chameides, B. (2005). *Global warming's increasingly visible impacts*. Environmental Defense.

Wewerinke-Singh, S. (2019). *State responsibility, climate change and human rights under international law*. Hart Publishing.

World Economic Forum. (2021). *Is the Paris Climate Agreement legally binding? Experts explain.* https://www.weforum.org/agenda/2021/11/paris-climate-agreement-legally-binding/

World Health Organisation. (2014). *Quantitative risk assessment of the effects of climate change on selected causes of death, 2030s and 2050s.*

World Meteorological Organisation. (2021, January). *Tropical cyclone Eloise hits Mozambique.* https://public.wmo.int/en/media/news/tropical-cyclone-eloise-hits-mozambique

World Vision. (2019). *2019 Cyclone Idai: Facts, FAQs, and how to help.* https://www.worldvision.org/disaster-relief-news-stories/2019-cyclone-idai-facts

CHAPTER 3

Climate Change Adaptation Governance in Africa: The Legal and Institutional Frameworks

Marus Gbomagba, Olubunmi A. Afinowi, and Naomi Kumi

1 Introduction

The adverse impacts of climate change are and will be more severely felt in Africa than in other regions of the world (African Ministerial Conference on Environment, 2014) despite the continent's negligible contribution to current and/or historic emissions. Ayompe et al. (2021, p. 3) pointed

M. Gbomagba (✉)
Research Institute of Environmental Law, School of Law, Wuhan University, Wuhan, China
e-mail: gbomagbamarus@gmail.com

O. A. Afinowi
Faculty of Law, Department of Private and Property Law, University of Lagos, Lagos, Nigeria

N. Kumi
Department of Atmospheric and Climate Science, University of Energy and Natural Resources (UENR), Sunyani, Ghana

M. Addaney et al. (eds.), *Climate Change in Africa*,
https://doi.org/10.1007/978-3-031-30050-9_3

out that over the period 1990–2017, Africa emitted only 3.8% of global greenhouse gas emissions, in contrast to 23% in China, 19% in the US, and 13% in the European Union. Similarly, Ritchie (2019) demonstrated that the continent has been responsible for less than 0.01% of all emissions over the last 266 years. From rising temperatures, rising sea levels, rainfall anomalies, and droughts to food insecurity, the people, economy, and the continent's ecosystems are paying a heavy price. Millions of lives are threatened across Sub-Saharan Africa due to the frequency and intensity of natural and climate disasters such as the devastating cyclones Idai and Kenneth, locust outbreaks in eastern Africa, and droughts in southern and eastern Africa (International Monetary Fund, 2020). Climate change is fundamentally unfair. The countries and populations most threatened by its impacts, and least able to adapt to them, are those that have contributed least to the problem. The threats posed by Climate Change compounded by other challenges such as, the Sahel's desertification, coupled with political instability and insecurity contribute to conflicts and mass migration on the continent (IMF, 2020; Rigaud et al., 2019).

To address the adverse effects of climate change, the international climate regime (the United Nations Framework Convention on Climate Change, its associated Kyoto Protocol, and the Paris Agreement) constitutes the framework at the global level within which adaptation and mitigation actions are undertaken. At the regional, sub-regional, and national levels, many initiatives are developed towards addressing the impacts of climate change in terms of adaptation strategies. At the African regional level, such initiatives include the African Ministerial Conference on Environment (AMCEN), the African Union Strategy on Climate Change, the Framework of Southern and Northern Africa Climate Change Programs, and the West and East African Community Climate Change Policy (African Ministerial Conference on Environment, 2014). In recent times, there are other initiatives, such as the Committee of African Heads of State and Government on Climate Change (CAHOSCC) and other sub-regional and sectoral policies and frameworks (AU Climate Change and Resilient Development Strategy and Action Plan). At the national level, some countries on the continent have also developed legal frameworks and strategies to address national climate change challenges. However, in a recent assessment, Salzinger & Desmidt (2021) finds that while many countries within the region have strategies for climate action, there is often a lack of alignment or regional coordination in terms of implementation. Although there are some laudable efforts, climate change adaptation governance on the continent remains a major problem for national decision-makers as remarked by the African Development Bank report on adaptation components in African NDCs (AfDB, 2019). The chapter addresses the lack of coordination of policies and institutions in Sub-Saharan Africa, which determined the current underwhelming state of adaptation governance on the continent. It also essays to point out the inadequacy of the policy framework for climate change adaptation governance. Thus, the guiding question is: How should policies be framed and coordinated to permit effective adaptation to climate change on the continent? Three countries (South Africa, Kenya, and Nigeria) serve as case studies to provide answers to this question.

2 Research Approach

Today, climate change is considered a global public issue and needs global answers (Pohlmann, 2011). The recent Intergovernmental Panel on Climate Change (IPCC) Special Report on the impacts of global warming of 1.5 °C showed the urgency for societies to accelerate adaptation to climate change (IPCC, 2022). Adaptation has consistently gained recognition on global policy agendas over the last few years (Persson, 2019). These issues continually lead to the need for innovations in climate change policies for countries in Africa. However, what choices and how in practice, adaptation should be governed for effective outcomes continue to be subject to debate (Huitema et al., 2016). Some scholars suggest that climate adaptation governance should be regarded as modes or mechanisms (Molenveld et al., 2020), while others refer to policy instruments that guide public action in responding to climate change's adverse impacts (Hall, 2011). There is also an increasing number of empirical studies that question the assumption that adaptation is (only) a local concern (Liverman, 2015). These studies discuss the global and transnational dimensions of adaptation (Persson, 2019). While most of these studies take a bottom-up perspective on governing adaptation (McLeman, 2010), many other studies examine climate change adaptation from a top-down perspective (Pohlmann, 2011). The process in the latter case is always seen to start from the highest level, namely the global regime, and then downward. A notable finding, which applies to African countries as well, is that climate change policies are a result of policy diffusion which can be attributed to internal drivers such as the slow rate of development (Jordan & Huitema, 2014).

Global adaptation governance often aligns with globally identified challenges and concerns and focuses on the implementation of international climate agreements, which often do not address local adaptation problems (Jordan & Huitema, 2014). This research acknowledges the potential role of global governance in strengthening climate adaptation in its various reports (Tan et al., 2022). It is also noted that since the adoption of the Paris Agreement, domestic climate governance has become a much more popular object of attention (Tan et al., 2022). While adaptation governance is not an ingrained field of governing when compared to the discussions on mitigation (Massey & Huitema, 2013), some interesting points stand out from a review of the literature. In the first place, adaptation is generally considered a matter of jurisdictional levels than mitigation (Huitema et al., 2016). This merely implies that mitigation tackles a global problem and is considered as needing an international and national approach. In contrast, adaptation is usually described as something that needs to occur at the local or regional level, perhaps with some facilitation or assistance from higher levels (Neufeldt et al., 2010). This consideration is not only relevant to specific local climate issues but also to varying local vulnerabilities and adaptive capacities. Secondly, because the substantial ideas in mitigation are specific and often quantifiable, e.g. the reduction of CO_2 emissions, in adaptation they are either broad (for instance, "to enhance adaptive capacity", "to increase resilience") or very domain-specific (for instance, "to maintain a flood protection standard") (Huitema et al., 2016). About this, it could be argued that climate change adaptation governance lacks a commonly agreed core aim. It is rather about everything without specific focus and as Wildavsky (1973) puts it, "if climate adaptation governance is everything, then maybe it is nothing?"

Additionally, climate adaptation governance requires the collective efforts of multiple societal actors to address problems or reap the benefits associated with the impacts of climate change (Huitema et al., 2016). It involves the creation of institutions, rules, and organizations, and the selection of normative principles to guide problem solutions and institution building (Persson, 2019). Therefore, actors involved in governing climate change adaptation should engage in making choices, for instance on problem definitions, jurisdictional levels, modes of governance, and policy instruments as climate change governance regimes evolve (Huitema et al., 2016). However, there is no evidence of how such choices affect the outcomes of efforts to govern adaptation (Persson, 2019). Thus, choices made and how issues are defined have important consequences for the support for governance interventions and their effectiveness. Likewise, another debate on adaptation governance more relevant in the African context is related to the questions of modes of governance, policy coordination, consensus building, and strategy formulation (Tan et al., 2022). The climate governance literature on the continent is unclear regarding how, policies should be formulated and implemented with coherence involving the different stakeholders (Sapiains et al., 2020). There is currently little relevant research on what is the current state of adaptation governance on the continent and how the integration of policies and programs across sectors and different actors could constitute an effective response to the climate crisis (Tan et al., 2022). There also remain large gaps in understanding of the state of adaptation in Africa generally (Ford et al., 2014). The assessment reports of the Intergovernmental Panel on Climate Change (IPCC, 2022) and other climate assessments (Costello et al., 2009), for instance, provide selected examples of adaptation in action, as do a variety of adaptation databases (e.g. Africa-adapt.net, weadapt.org), but do not systematically and comprehensively compile and assess information on current actions (Ford et al., 2014). Other studies examine adaptations supported by specific funding schemes but provide only snapshots of broader trends (Ford et al., 2014). These challenges on the African continent have been attributed to weak institutions and uncoordinated action at various levels of governance (African Union Climate Change Action Plan, 2022–2032).

This chapter is a review of databases, grey literature sources, and the Web of Science for articles that address climate change adaptation in the African Union context with key countries' cases. Relevant grey literature was identified using two focused searches on Google, focusing on Africa and the case studies countries. Literature located through searches of article references or targeted internet searches helped identify the relevant works on the adaptation topic included in the review. Extracted data on the characteristics of the included articles and books including their study design and outcome measure and, where possible, commonalities or differences between the study findings were attempted for highlighting. Policy documents were located by searching the websites of the Adaptation Policies of different countries' cases and the common position at the African Union level. This chapter is an assessment of knowledge and practices on adaptation governance on the continent within the framework initiatives of the United Nations Framework Convention on Climate

Change (UNFCCC), the Intergovernmental Panel on Climate Change (IPCC), the Kyoto Protocol, and the Paris Agreement among others at the global level. Regional initiatives at the African Union and the Regional Economic Communities (RECs) actions, together with national adaptation policies and climate laws are explored to derive the inconsistency of actions.

3 The Conceptualization of Climate Adaptation

Adaptation issues significantly gained recognition in international climate politics despite the attention tentatively turned towards mitigation in the post-Paris discussions (Hall & Persson, 2017). Firstly, discussed in terms of the developing countries issue, it is now considered a global public discourse (Juhola & Westerhoff, 2011). Over the past two decades, the challenge of climate change adaptation has emerged as a complex set of issues facing public policymakers (Bednar et al., 2018). The most common definition of adaptation can be found in the Intergovernmental Panel on Climate Change's IPCC (2022) most recent assessment report which provides that:

> Adaptation is defined, in human systems, as the process of adjustment to actual or expected climate and its effects to moderate harm or take advantage of beneficial opportunities. In natural systems, adaptation is the process of adjustment to actual climate and its effects; human intervention may facilitate this.

Considering this definition, adaptation thus includes both anticipatory and reactive actions to reduce vulnerability or enhance resilience in response to already observed and expected changes in climate. Adaptation practices, for example, can be differentiated along several dimensions, including the spatial scale, sectors, actions, or actors (Juhola & Westerhoff, 2011). The Africa Climate Change Strategy (2020) highlights that adaptation to the impacts of climate change will remain the universal top priority action for Africa and a prerequisite for building resilience. This is confirmed by the NDCs submitted by African countries where adaptation is the focus (African Union, 2020). It is worth recalling that the adverse impacts of climate change differ according to the location (du Plessis & Kotzé, 2014), hence specific adaptation actions by the stakeholders will be determined by the threats they face. For instance, it is projected that spatially extensive droughts and the most extensive projected decreases in rainfall would happen in northern Africa and the southwestern parts of South Africa. In contrast, there will be an increase in precipitation in East Africa (African Union, 2020). This shows that each sub-region of Africa is and will be impacted differently, further adding complexity to the crisis and requiring different scaled adaptation measures.

4 Defining Governance in Climate Change Adaptation

It is widely accepted today that climate actions cannot only focus on mitigation but the need for society to adapt to the impacts of climate change is pressing (Peltonen et al., 2010). Addressing climate change adaptation means addressing significant governance issues (Termeer et al., 2017). Adaptation to climate change is then not only a complex technical issue, but it is also a demanding matter of governance that requires in-depth understanding and planning to achieve tangible results on the ground. Climate adaptation governance implies different framing of the matter, different levels of governance (national, regional, and global), and different actors (public, civil society, and local administrators). It is also linked to countries' development, and the well-being of nature and people as governance is about making decisions (Caffrey & Farmer, 2014), which state actors are more equipped to carry out due to their ability to employ measures for both implementation and enforcement (Jordan & Huitema, 2014).

According to Bovaird and Löffler (2009, p. 8), "trying to define public governance seems to open Pandora's box" due to the complexities of governance in the public space. However, they provide that governance may be seen as how an organization works with its partners, stakeholders, and networks to influence the outcomes of public policies (p. 6). Termeer et al. (2017), suggest that governance entails interactions between public and/or/private entities ultimately aiming to realize collective goals. Caffrey and Farmer (2014) in their part, see governance as the mechanism that revolves around structures and processes for sharing and using the power that shapes decisions. From this broad perspective, governance includes the domains of operations and management, fiscal policy, planning, budgeting, rule of law, regulation, discursive debate, negotiation, mediation, conflict resolution, elections, civic engagement, and other formal and informal decision-making processes (Caffrey & Farmer, 2014).

Governing climate adaptation could then mean bringing together multiple societal actors and stakeholders in their respective efforts to address problems or reap the benefits, associated with the impacts of climate change. In other words, climate adaptation governance involves the creation of institutions, rules, and organizations, and the selection of normative principles to guide problem solutions and institution building (Molenveld et al., 2020). This multi-level character of climate adaptation governance poses some specific, demanding challenges, such as the context of institutional fragmentation, as climate change involves almost all policy domains, sectors, and governance levels. There are also persistent uncertainties about the nature and scale of risks and proposed solutions; and the need to make short-term policies based on long-term projections (Persson, 2019). The multi-level nature of governance, therefore, requires the strengthening and promotion of innovation, problem-solving capacity, learning, and the development of solutions that benefit more sectors.

As discussed above, the fundamental commonality in defining governance within the climate adaptation context is the process of including different stakeholders who, in some form of participatory agreement, deal with the causes and consequences of climate change (Pohlmann, 2011). Literature on global climate governance which is initially predominated by a single regime (the UNFCCC, its associated protocol, and agreement with a focus on state actors) is moving towards a more complex arrangement (Persson, 2019). The shift in the global regime is now towards networked climate governance (Bäckstrand, 2008), transnational climate governance, and polycentric governance (Bansard et al., 2017). In this light, it is noted that the diffusion of policies across states or levels depends on factors including trialability, the low complexity and suitability to tackling visible problems which are often attributable to climate adaptation Jordan and Huitema (2014)

In Africa, the tendency is also a complex regime with non-state climate actions, international cooperative initiatives, and national-level government policy (Averchenkova et al., 2019). However, while there is a plethora of stakeholders, state actors are essential in the innovation and diffusion of policies due to their control of the institutional, administrative, and legal framework for climate adaptation (Jordan & Huiteman, 2014). The African Ministerial Conference on Environment, 2014 defines climate governance as the exercise of power and authority by formal institutions of governments to minimize the impacts of climate change on communities, ecosystems, and the wider environment in general. The strategy also provides for the obligation of developing legislation, policies, and institutional and management frameworks, at continental, regional, and national levels (African Ministerial Conference on Environment, 2014). Specifically, the AU Climate Change Strategy and Action Plan 2022–2032 identifies legislation as playing a vital role in influencing the success of adaptation initiatives (p. 37). The next section of the chapter looks at the regulatory framework of climate adaptationon the African continent. This includes existing legal, institutional, and policies mechanism at the international and regional levels in the relevant countries under consideration.

5 Climate Adaptation Governance in Africa: the Regulatory Framework

"Climate change regulatory framework" is used in this section to describe the legal, policy, and institutional frameworks on adaptation in Africa. This is because the nature of international and national instruments on climate change is incapable of being strictly classified as "laws" or "policies" (Jegede, 2016, p. 14). These instruments emerge in the context of intersecting governance as explained in the previous section of this work. According to Jegede (2016, p. 16), the regulatory framework on climate governance emanates from international climate negotiations between States under international law. These instruments, which include treaties such as the UNFCCC, the Kyoto Protocol, the Paris Agreement, and the decisions of the Conference of Parties (COP), constitute the legal regime that governs climate actions on the international stage. Other relevant guidelines include the decisions of the Meeting of Parties (MOP), as well as reports from different UN organs although they may not qualify as a set of law, influence state conduct over time (Jegede, 2016, p. 17). The international framework on climate change is further adopted by individual states, including those under consideration and forms the basis of national and sub-national climate action and governance. Also, they frame how climate change could and should be governed across levels (Jernnäs, 2021).

In line with the multi-level nature of climate action, the development of a climate change adaptation regulatory framework occurs in and across levels and tiers of governance. In this regard, at the African regional level, there are also efforts of strategies, policies, and institutional frameworks in line with the UNFCCC regime (Sapiains et al., 2020). As noted earlier, internal drivers play a significant role in driving adaptation policies within the Region, to this end, there exist various policies and strategies which have been aimed at aiding Africa's adaptation to climate impacts (heat waves, drought, flood, and fire) for years. The advancement of Africa's common position on climate change is based on initiatives at the AU level (Addaney, 2018). To this end, the overall objective of the AU Climate Change Strategy and Action Plan, 2022–2032 is to build resilience in African communities and support regional adaptation. As outlined in the 2022–2032 AU Action Plan, it is imperative that adaptation follows a path of inclusion, cooperation, and joint ownership of climate action, policies, plans, and strategies across all levels of governance and stakeholders (p. 7). This includes a coordinated collaboration between state actors at the sub-regional level, in this regard, the 8th ordinary AU Assembly session urged member states and the Regional Economic Communities (RECs) to include climate change in their respective development policies and programs (Addaney, 2018).

In recognition of the low level of climate-related research within the Continent (IPCC Sixth Assessment Report, Fact Sheet -Africa), the AU 2022–2032 Action Plan requires that climate action within the Continent must align with Agenda 2063 and be based on research within the Continent. While this may seem somewhat ambitious, it is worth pursuing in order to ensure policies and actions that cater to the socio-economic and cultural peculiarities of the African continent. Related institutions within the framework of the AU that play a pivotal role in climate adaptation governance in Africa, are the Conference of African Heads of State and Government on Climate Change (CAHOSCC) and the African Ministerial Conference on the Environment (AMCEN). Other inter-governmental actors involved in the policy, institutional, and management frameworks, at continental, and regional levels include the United Nations Economic Commission for Africa (UN ECA), the African Union Commission (AUC), the African Development Bank (AfDB), and Regional Economic Communities (RECs).

6 National Regulatory Frameworks on Climate Adaptation

At the national level, many countries on the continent have adopted the international and regional frameworks on climate change adaptation in their domestic policies, and laws, and developed new legal and institutional frameworks to implement them. African countries have been active in the global climate change negotiations and are all Parties to the UNFCCC, the Kyoto Protocol, and the Paris Agreement of 2015 with many countries possessing various policies and structures evidencing climate action, commitments, and strategies. According to Averchenkova's et al. assessment (2019), South Africa has put in place one of the most elaborate and consultative climate governance systems observable among developing and emerging economies, significant aspects of this are the country's National Climate Change Response Strategy, followed by the 2011 National Climate Change Response White Paper (NCCRWP). In 2012 climate change became a key element of the National Development Plan, the overarching plan for the country (du Plessis & Kotzé, 2014). South Africa also currently has a Climate Change Bill under consideration, the aim being to legislate upon the country's key strategic climate change objectives, governance elements, and policies (Government Gazette, 2022). It is noteworthy that the National Climate Change Adaptation Strategy of 2020 has helped to put policies on adaptation and resilience in proper perspective and direct adaptation action in South Africa (Averchenkova et al., 2019). The Strategy currently defines the country's vulnerabilities, plans to reduce those vulnerabilities and leverage opportunities, and outlines the required resources for such action, while demonstrating progress on climate change adaptation (Government Gazette, 2022). In addition to a robust policy framework, South Africa also established an Inter-governmental Committee on Climate Change, the purpose of which is to engender inter-sectoral and multi-level climate governance across relevant ministries and departments of government (Taylor, Cartwright & Sutherland, 2014).

Kenya is one of the few countries and the first on the continent with a substantive climate change law (Dlamini et al., 2021). The country enacted the Climate Change Act in May 2016, after several years of political negotiation, including a presidential veto in 2013. The Act applies across the economy and sub-national levels with the overall intent of providing mechanisms to build resilience and low-carbon development (Worker, 2017). It also contains some relevant provisions on adaptation mainstreaming. For instance, Section 3(2) of the Act establishes the obligation to mainstream climate responses into development planning, decision-making, and implementation (Climate Change Act, 2016).

The Act also provides obligations to build resilience and enhance adaptive capacity to the impacts of climate change; formulate programs and plans to enhance the resilience and adaptive capacity of human and ecological systems to the impacts of climate change (Section 3(2)(a)). Further, it requires mainstreaming and reinforcing climate change disaster risk reduction into strategies and actions of public and private entities (Section 3(2)(d)). Finally, it also establishes a Climate Change Directorate to implement the law, enforce compliance, and coordinate activities (Section. 9). In recognition of the necessity for multi-scalar adaptation, the Act imposes specific duties on both public and private entities, as well as mainstreams climate action into strategic sectors (Part IV, Climate Act, 2016). It is also worth noting that the legislation makes provisions for public participation and access to information (Section. 24). It is noted here that public participation and access to information pertaining to government action on climate adaptation could provide a means for stakeholders to evaluate government action (Jordan & Huitema, 2014). Also fundamental, is the implementation of the Act by both levels of government within the country, that is, the national government and the county governments. While the national government has the responsibility to formulate and review climate policies, the county governments are required to integrate the national policies into their County Integrated Development Plans across sectors (Clarice Wambua, 2019).

Besides the Climate Change Act, the Republic of Kenya has a robust climate change policy framework and a sophisticated system of domestic institutions (Bellali, 2018). This policy framework includes the National Climate Change Response Strategy, (NCCRS 2010), the National Climate Change Action Plan, (NCCAP 2013–17), reviewed for 2018–2022, the National Adaptation Plan, NAP (2015–2030), the National Climate Change Framework Policy (2016), and the Green Economy Strategy and Implementation Plan (GESIP 2016–2030) (Bellali,2018). This broad range of policies aims to foster low-emission development and enhance climate resilience. It is noted that the Kenyan Climate Change Act adopts a multi-level climate governance framework which makes climate action more effective, and this approach seems to have engendered climate adaptation at the county levels within the country (Clarice Wambua, 2019).

Nigeria has also passed a climate change law known as the Climate Change Act of 2021. The Act is the flagship legislation on climate action in the country. Similar to the Kenyan Act, the federal government is tasked with the responsibility of setting guidelines for and monitoring climate action within the country, while government ministries, departments, and agencies across other tiers of government are required to take necessary action in line with the provisions of the Act (Sections 1 and 2). The Act also makes provisions for public engagement (Section 30) and stipulates stakeholders' obligations in relation to climate action within the country (Part VI). The National Climate Change Policy was developed in 2013 and it defines strategies for building a climate-resilient society, and promoting low-carbon, high-economic growth in key sectors (energy, mining, industry, population and settlements, health, tourism, and transport). This Policy also highlights other climate-related policies, such as the 2007 National Policy on Drought and Desertification (Dlamini et al., 2021).

There are also several other adaptation-related policies such as the National Adaptation Strategy and Plan of Action on Climate Change for Nigeria, the National Agricultural Resilience Framework Nigeria, the National Integrated Infrastructure Master Plan, and the National Action Plan on Gender and Climate Change for Nigeria (Dlamini et al., 2021). These policies seek to provide effective mechanisms for climate adaptation to foster a climate-resilient economy (Afinowi, 2020). Nigeria's 2021 Climate Change Act provides an ambitious framework for mainstreaming climate actions in line with national development priorities and sets a net-zero target for 2050–2070 (Section 1). It also provides a solid framework for climate action to achieve Nigeria's short-, medium-, and long-term goals for climate adaptation (Ladan, 2022). While there will be several necessary measures to operationalize Nigeria's 2021 Climate Change Act, it demonstrates the country's seriousness in approaching climate action. Notably, quite unlike the Kenyan Act, the Nigerian Climate Change Act skirts the issue of collaborative governance across tiers of government, especially the state governments; this seems like a missed opportunity to streamline climate action and strategies across the country.

South Africa has undoubtedly blazed the trail in innovations in climate change adaptation in Africa, this is despite not having passed the Climate Change Bill into law yet. Kenya also occupies a notable spot with research showing that the country had the highest number of documented adaptation initiatives in Africa as of 2015 (Ford et al., 2015). The existing policies and institutional framework provide a viable basis for adaptation. Further, multi-level climate collaboration in the country has built on the inter-governmental collaboration instituted in the country's Constitution

(Constitution of the Republic of South Africa (Act No. 108 of 1996)). Kenya is also a pioneer in establishing a legislative framework. There is thus the likelihood that other countries within the continent will emulate both countries based either on the measurable "beneficial outcome" of their processes or projected improvements from the status quo Jordan and Huitema (2014) .

7 Governance Barriers to Adaptation in Africa

Discussions in the previous section of this chapter show the existence of a regulatory framework at the AU level with strategies put in place to guide national adaptation actions. However, this regulatory framework does not allow the successful delivery of adaptation measures to climate change on the continent (du Plessis & Kotzé, 2014). A major bane to adaptation in Africa is the absence of a clear regional governance policy. As noted, governance issues constitute significant barriers that hamper the development of effective and timely adaptation responses (Jegede, 2016). As observed by Alam et al. (2019, p. 33), the lack of proper implementation of climate policies and laws coupled with weak coordination between regional and national institutions or agencies are key barriers to the implementation of adaptation measures. For instance, the AU Strategy and Action Plan (2022–2032) does not adequately address the novel policy concerns presented by the scale of adaptation which require new forms of decision-making processes. Furthermore, as the AU has not yet adopted any substantive regional treaty on climate change adaptation (Addaney, 2018), there are no binding obligations on state members who fail to align with the regional framework. This implies that adaptation governance efforts from the AU level like the international level appear to have been intentional but not particularly authoritative (Persson, 2019). Moreover, the fact that at the international level, there is a wide disparity in the policy process between the least developed, developing, and developed countries (Alam et al., 2019, p. 33) impedes the implementation of effective adaptation measures.

Though South Africa has taken a strong stance on climate change on the international stage and has developed a significant and forward-looking set of policies, key governance challenges alter adaptation results (Averchenkova et al., 2019). The current law and policy framework are disjointed and patchy at best and there is still no coherent and explicit outline of what each stakeholder can and should do in the climate adaptation governance context (du Plessis & Kotzé, 2014). It can be argued that the strategic intent of most policies is good, but implementation mechanisms in many cases are not sufficiently developed or structured to meet the pressing needs of adaptation (Averchenkova et al., 2019). While discussing key governance challenges for the implementation of adaptation strategies in South Africa, du Plessis and Kotzé (2014) and Averchenkova et al. (2019) identify some major barriers. These major barriers are the lack of policy alignment, coherence, and coordination; policy complexity and continuity over time; limited staff capacity and financial resources; and issues with public–private engagement and consultation (Averchenkova et al., 2019). The lack of policy alignment across sectors and tiers of climate governance, as noted by du Plessis and Kotzé (2014) explains the institutional inertia observed towards effective implementation of adaptation measures. Ineffective coordination on climate change horizontally among national government departments and vertically between national, provincial, and municipal governments, as well as with non-state actors, is highlighted as a critical barrier to effective adaptation. For example, there is no coherent and explicit outline of what municipalities can and should do in the implementation of adaptation strategies (du Plessis & Kotzé, 2014). This suggests that there is a lack of high-level direction and clear institutional mandates for implementation.

A perceived lack of capacity to deal with climate change adaptation and related policies within the South African government and sectoral departments is another major governance challenge (Averchenkova et al., 2019). This limitation may stem from limited human and financial resources to a shortage of relevant expertise and skills. In addition to limited human resources, there is growing complexity of work involved in designing and implementing sectoral and multi-sector resilience policies. Technical skills and capacity to work on climate change are in even greater shortage for many provincial governments and municipalities than they are for central government departments, as noted by Averchenkova et al. (2019). Shortage of adequate financial resources is among the top barriers to effective governance and implementation of climate adaptation policies in South Africa, particularly at the municipal level (Yelland, 2016). One of the critical gaps underlying the shortage of finance mentioned by the experts was the lack of a comprehensive climate finance strategy to define the allocation of resources to support climate adaptation work and develop a plan for attracting international funding and investment (Averchenkova et al., 2017). Finally, governance issue hampering adaptation implementation is the mistrust of public–private engagement (Averchenkova et al., 2019). A profound mistrust between the various players renders it difficult to deliver adaptation results (Averchenkova et al., 2019).

Kenya, despite being the first country in Africa to have adopted clear legislation on climate change, still faces governance issues in implementing adaptation measures (Dlamini et al., 2021). Despite its adoption 6 years ago, the Climate Act is yet to be fully operationalized to deliver adaptation responses in practice (Bellali, 2018). For example, many of the institutions such as the National Climate Change Council (NCCC), established by the Act had, at the time of writing this work, not been constituted despite Section 35 of the Act requiring members of the Council to be appointed within three months after the Act enters into force (Government of Kenya, 2016). The Council has therefore not yet met, and the delay has affected the operationalization of the Act, including the Climate Change Fund that could help in implementing adaptation measures (Dlamini et al., 2021). Sections 22 and 36 of the Climate Change Act require the Environment Cabinet Secretary to develop subsidiary legislation for the effective implementation and operationalization of the Act; these regulations are yet to be developed (Worker, 2017).

The lack of clear timelines for enacting regulations or strategies, including ways to enforce timelines; establishing clear dispute settlement procedures; and the need for broad (state and non-state) stakeholder engagement hinders the overall application of the act (Dlamini et al., 2021). Another relevant constraint in the implementation of the Act at the county's level to favour adaptation is the inadequacy of the amount of action planning and resource allocation, which have even resulted in maladaptation (Ojwang et al., 2017). Additionally, counties do not have jurisdiction over some sectors relevant to climate action, such as water resources management, wildlife, and mining; these functions remain with national departments like the Water Resources Authority (Bellali, 2018). Kenya's multi-level climate policy system is still challenged by governance barriers with vertical coordination hindered by unclear and overlapping mandates (Bellali, 2018). Planning frameworks are not well aligned across governance levels and do not result in climate proofed investments (Mwangi & Mutua, 2015).

Like the two previous cases discussed above, there is currently no statutory coordination mechanism for the myriad of national policies, programs, and plans to ensure cohesion and effective tracking and monitoring of adaptation implementation in Nigeria (Dlamini et al., 2021). The Act does not provide for inter-governmental collaboration or multilevel collaboration in climate action. Amobi and Onyishi (2015) note that adaptation strategies and plans have suffered more from institutional and structural bottlenecks than from the will to embrace the problem in the first place. The development of adaptation policies and strategies is hardly documented by state actors (Dlamini et al., 2021). Government institutions are faced with major challenges that undermine adaptive capacity, ranging from requisite resources to low-level human capital (Amobi & Onyishi, 2015). There is also inadequate investment in strategic areas for climate change adaptation coordination (Moram, 2011). In addition, there is limited space for civic engagement (particularly for NGOs & CBOs) due to financial, human resource, and political constraints (Amobi & Onyishi, 2015). In addition to the above, the country possesses an abysmal track record in maintaining regulatory standards of environmental protection, it would seem that this apathy extends to climate action.

The above general factors militating against climate change adaptation policies and programs can be added to the built institutional problems of poor accountability, financial corruption, and the disruption of political processes due to regime instability in the Nigerian polity (Igwe & Fukuoka, 2010). A recent audit of the Ecological Fund (a voluminous federal fund for undertaking preventative and remedial environmental projects) shows that civil servants and contractors have connived to embezzle billions of dollars since the establishment of the Fund in 1981 (Page, 2016). Furthermore, even though the government has spent $14 billion since 1999 on modernizing the power sector, the national electricity supply steadily diminished to a level on par with the city of Edinburgh, which has a population of 500,000 (Page, 2016). The latest Transparency International (2022) Corruption Perceptions Index (CPI) report shows the country scoring 24 out of 100 points and ranked 154 out of 180 countries surveyed. It demonstrated the country falling back five places from the rank of 149 in 2020 and placing as the second most corrupt country in West Africa (Transparency International, 2022). While Nigeria's 2021 Climate Change Act provides an ambitious framework for mainstreaming climate adaptation actions, several necessary measures are needed to operationalize it (Ladan, 2022). Regarding its implementation, it is too early to have a fair assessment of how it has translated into adaptation outcomes.

Based on the findings above, this chapter offers some key recommendations for the effective implementation of adaptation measures in the case countries. These recommendations are also relevant for other countries across the continent. At the continental level, the AU should adopt a substantive regional framework convention to govern adaptation and provide binding guidelines on how adaptation can be mainstreamed into national adaptation policies and strategies. There should also be a mechanism to follow up on the implementation of existing strategies in member states. The formulation, implementation, and monitoring of adaptation policies and plans should be participatory and inclusive in a top-down approach. As South Africa moves towards the adoption of a framework law on climate change (Government Gazette, 2022), the country should align development policies and strategies with the objectives of the various adaptation policies.

The successful implementation of adaptation measures requires renewed political commitment and leadership from the highest level and a unified approach from the Government. There needs to be a clear mandate for a lead agency entrusted with coordinating implementationand for each of the sectoral agencies to designate staff and resources and implement policies, based on the existing work and lessons learned from past experiences (Averchenkova et al., 2019). There is a need to move from climate change policy development to implementation not only in South Africa but across the continent. This requires a transparent and continuous process focused on policy alignment and coordination of implementation, led by a strong government agency that has a clear mandate from the highest political level to coordinate this process. The key to leveraging buy-in and effective participation across levels of governance, horizontally and vertically, is to link climate adaptation policy to poverty reduction objectives and clean energy availability for communities.

Key among other recommendations will be ensuring that there is buy-in from all relevant stakeholders, from within and outside government. This can help mobilize resources to support the drafting and adoption of the climate framework law as was the case in Kenya and Nigeria. As Kenya and Nigeria learned, the failure to engage with a broad range of stakeholders and build consensus resulted in failed attempts to legislate.

Applying the law once it is enacted can be equally difficult, as Kenya's example illustrates. Harnessing political will to give effect to the provisions of the legislation is therefore key. Including time frames for specific actions in the framework law may also help to ensure that it does not just remain law in the statute book. Where institutions are fit for purpose or can be adapted to cover climate change policies by clarifying or expanding their mandate, countries may opt for this as it will help minimize funding requirements and organizational reform. In some cases (such as Kenya) it may be necessary to first streamline the institutional framework and clarify the role/mandate, if any, of various existing agencies and new institutions.

8 Conclusion

This chapter has reviewed the status of adaptation governance in Africa concerning the existing regulatory framework and governance barriers delaying effective climate action. It assessed that, though a framework climate binding regulation has not yet been enacted at the continental level, solid guidelines do exist. In the same way, many African States have viable guiding policies and action plans and some even adopt climate laws (as is the case in Kenya and Nigeria). The review, however, pointed out that despite the plethora of policies, strategies, and some laws addressing climate adaptation, governance issues still hinder effective adaptation measures. The chapter demonstrated that key governance barriers on the continent hampering adaptation is the lack of policy alignment, coordination, and inadequacy of technical capacity and poor climate financing. It then suggests that adaptation measures are formulated and implemented moving to a more synergistic approach. It also highlights that a multi-tier approach is required to build the capacities of governments and communities in Africa to effectively adapt to climate change. Monitoring and reporting are also identified as strategies that could promote effectiveness in adaptation governance. To this end, the chapter opines that it is essential that countries' regulatory frameworks have clearly defined mandates/functions on policy formulation, coordination, implementation, advice, and oversight/monitoring to avoid redundant policies and legislation. While the research recognizes the importance of state actors, climate adaptation must entail a viable interaction process between state and non-state actors within the national and sub-national space as part of key measures to improve the effectiveness of climate governance. This would help overcome obstacles in horizontal and vertical coordination and stakeholder engagement.

References

Addaney, M. (2018). Climate change adaptation law and policy in the African Union: Creating legal pathways for adaptation. In *Conference Proceedings of Adaptation Futures 2018*. https://openbooks.uct.ac.za/uct/login?source=%2Fuct%2Fcatalog%2Fbook%2FAF18. Accessed 14 May 2022.

African Development Bank (AfDB). (2019). *African Development Bank—building today, a better Africa tomorrow*. https://www.afdb.org/fileadmin/uploads/afdb/Documents/Generic-Documents/Analysis_of_Adaptation_Components_in_African_NDCs_2019.pdf. Accessed 22 May 2022.

Afinowi, O. (2020). Climate change adaptation and its implementation at national and international levels. *Journal of Sustainable Development Law and Policy (The), 11*(1), 138–159. https://doi.org/10.4314/jsdlp.v11i2.7. Accessed 15 May 2022.

African Ministerial Conference on Environment (AMCEN). (2014). *African strategy on climate change*. AMCEN-15-REF-1.

African Union. (2020). Draft Africa Climate Change Strategy 2020–2030. Addis Ababa, Ethiopia: the AU. https://archive.uneca.org/sites/default/files/uploaded-documents/ACPC/2020/africa_climate_change_strategy_-_revised_draft_16.10.2020.pdf

African Union Climate Change and Resilient Development Strategy and Action Plan (2022–2032).

Alam, S., Gebremichael, M., Li, R., Dozier, J., & Lettenmaier, D. P. (2019). Climate change impacts on groundwater storage in the Central Valley, California. *Climatic Change, 157*(3-4), 387–406.

Amobi, D., & Onyishi, T. (2015). Governance and climate change in Nigeria: A public policy perspective. *Journal of Policy and Development Studies, 9*(2), 199–209.

Averchenkova, A., Fankhauser, S., & Nachmany, M. (Eds.). (2017). *Trends in climate change legislation*. Edward Elgar Publishing.

Averchenkova, A., Gannon, K. E., & Curran, P. (2019). *Governance of climate change policy: A case study of South Africa* (pp. 1–39). Grantham Research Institute on Climate Change and the Environment Policy Report.

Ayompe, L., Davis, S., & Egoh, B. (2021). *ShieldSquare captcha*. Iopscience.iop.org. https://iopscience.iop.org/article/10.1088/1748-9326/abc64f. Accessed 14 May 2022.

Bäckstrand, K. (2008). Accountability of networked climate governance: The rise of transnational climate partnerships. *Global Environmental Politics, 8*(3), 74–102.

Bansard, J. S., Pattberg, P. H., & Widerberg, O. (2017). Cities to the rescue? Assessing the performance of transnational municipal networks in global climate governance. *International Environmental Agreements: Politics, Law and Economics, 17*, 229–246.

Bednar, D., Raikes, J., & McBean, G. (2018). *The governance of climate change adaptation in Canada*. Institute for Catastrophic Loss Reduction.

Bellali, J. (2018). *Multi-level climate governance in Kenya. Activating mechanisms for climate action.*

Bovaird, T., & Löffler, E. (2009). Understanding public management and governance. *Public Management and Governance, 2*, 3–13.

Caffrey, P., & Farmer, A. (2014). *Adaptation and governance*. Pdf.usaid.gov. https://pdf.usaid.gov/pdf_docs/PA00K67X.pdf. Accessed 19 May 2022.

Costello, A., Abbas, M., Allen, A., Ball, S., Bell, S., Bellamy, R., Friel, S., Groce, N., Johnson, A., Kett, M., & Lee, M. (2009). Managing the health effects of climate change: lancet and University College London Institute for Global Health Commission. *The lancet, 373*(9676), 1693–1733.

Dlamini, C., Wabwoto, E., Mshelia, H., Chidia, N., Maduekwe, B. N., & Bird, P. (2021). *European capacity building initiative*. Available at https://www.ecbi.org

du Plessis, A., & Kotzé, L. (2014). The heat is on: Local government and climate governance in South Africa. *Journal of African Law, 58*(1), 145–174. https://doi.org/10.1017/S0021855314000047. Accessed 16 May 2022.

Ford, J. D., Berrang-Ford, L., Bunce, A., McKay, C., Irwin, M., & Pearce, T. (2015). The status of climate change adaptation in Africa and Asia. *Regional Environmental Change, 2015*(15), 801–814.

Ford, J. D., McDowell, G., & Jones, J. (2014). The state of climate change adaptation in the Arctic. *Environmental Research Letters, 9*(10), 104005.

Government Gazette. (2022). No. 41689 Vol. 636 8. Climate Change Bill (B9–2022). June 2022 at https://www.parliament.gov.za/bills. Accessed 3 June 2022.

Government of Kenya. (2016). Environment.go.ke. 2016. Accessed 2 June 2022)

Hall, N., & Persson, Å. (2017). Global climate adaptation governance: Why is it not legally binding? *European Journal of International Relations, 24*(3), 540–566. https://journals.sagepub.com/doi/full/10.1177/1354066117725157. Accessed 19 December 2021.

Hall, N. (2011). *Climate change and institutional change in UNHCR. Climate change and migration: Rethinking policies for adaptation and disaster risk reduction*, 102–114.

Huitema, D., Adger, W., Berkhout, F., Massey, E., Mazmanian, D., Munaretto, S., Plummer, R., & Termeer, C. (2016). The governance of adaptation: Choices, reasons, and effects. Introduction to the special feature. *Ecology and Society, 21*(3). https://www.ecologyandsociety.org/vol21/iss3/art37/. Accessed 18 December 2021.

Igwe, O., & Fukuoka, H. (2010). Environmental and socio-economic impact of erosion in Nigeria, West Africa. *International Journal of Erosion Control Engineering, 3*(1), 102–109.

International Monetary Fund (IMF). (2020). *Adapting to climate change in Sub-Saharan Africa*. https://doi.org/10.5089/9781513536835.086. Accessed 13 March 2022.

Ipcc.ch. (2022). *IPCC sixth assessment report*. https://www.ipcc.ch/report/ar6/wg2/. Accessed 16 May 2022.

Jegede, A. (2016). *The climate change regulatory framework and indigenous peoples' lands in Africa: Human rights implications*. Pulp.up.ac.za. https://www.pulp.up.ac.za/monographs/the-climate-change-regulatory-framework-and-indigenous-peoples-lands-in-africa-human-rights-implications. Accessed 17 May 2022.

Jernnäs, M. (2021). *Governing climate change under the Paris regime: Meeting urgency with voluntarism* (Doctoral dissertation). Linköping University Electronic Press.

Jordan, A., & Huitema, D. (2014). Innovations in climate policy: Conclusions and new directions. *Environmental Politics, 23*(5), 906–925.

Juhola, S., & Westerhoff, L. (2011). Challenges of adaptation to climate change across multiple scales: A case study of network governance in two European countries. *Environmental Science & Policy, 14*(3), 239–247.

Kenyan Climate Change Act No 11 of 2016.

Ladan, M. (2022). *Nigeria's climate change act and policy 2021 and the future of climate litigation*. Available at SSRN https://doi.org/10.2139/ssrn.4019698

Liverman, D. (2015). Reading climate change and climate governance as political ecologies. In *The Routledge handbook of political ecology* (pp. 303–319). Taylor and Francis.

Massey, E., & Huitema, D. (2013). The emergence of climate change adaptation as a policy field: The case of England. *Regional Environmental Change, 13*(2), 341–352.

McLeman, R. (2010). Impacts of population change on vulnerability and the capacity to adapt to climate change and variability: A typology based on lessons from "a hard country". *Population and Environment, 31*(5), 286–316. https://doi.org/10.1007/s11111-009-0087-z. Accessed 19 December 2021.

Molenveld, A., van Buuren, A., & Ellen, G. (2020). Governance of climate adaptation, which mode? An exploration of stakeholder viewpoints on how to organize adaptation. *Climatic Change, 162*(2), 233–254. https://link.springer.com/article/10.1007/s10584-020-02683-9. Accessed 15 November 2021.

Moram, C., Tittensor, D. P., Adl, S., Simpson, A. G. B., & Worm, B. (2011). How many species are there on earth and in the ocean? *PLoS Biology, 9*, 1100–1127.

Mwangi, K. K., & Mutua, F. (2015). Modeling Kenya's vulnerability to climate change–A multifactor approach. *International Journal of Science and Research, 6*, 12–19.

Neufeldt, H., Jochem, H., Hinkel, J., Huitema, D., Massey, E., Watkiss, P., Mcevoy, D., Rayner, T., Hof, A., Lonsdale, K., & Barker, T. (2010). Climate policy and interlinkages between adaptation and mitigation. In *Making climate change work for us: European perspectives on adaptation and mitigation strategies* (pp. 3–30). Cambridge University Press.

Nigerian Climate Change Act of 2021.

Ojwang, L., Rosendo, S., Celliers, L., Obura, D., Muiti, A., Kamula, J., & Mwangi, M. (2017). Assessment of coastal governance for climate change adaptation in Kenya. *Earth's Future, 5*(11), 1119–1132.

Page, M. (2016). *Corruption Hampers Nigeria's efforts to fight climate change*. Worldpoliticsreview.com. https://www.worldpoliticsreview.com/trend-lines/20339/corruption-hampers-nigeria-s-efforts-to-fight-climate-change. Accessed 31 May 2022.

Peltonen, L., Juhola, S., & Schuster, P. (2010). *Governance of climate change adaptation: Policy review*. http://www.baltcica.org/documents/baltcica_policyreview_090310_SJ.pdf. Accessed 25 December 2021.

Persson, Å. (2019). Global adaptation governance: An emerging but contested domain. *WIREs Climate Change, 10*(6). https://wires.onlinelibrary.wiley.com/doi/full/10.1002/wcc.618. Accessed 9 December 2021

Pohlmann, A. (2011). Local climate change governance. In A. Engels (Ed.), *Global transformations towards a low carbon society* (5, Working Paper Series). University of Hamburg/Klima Campus.

Rigaud, K. K., de Sherbinin, A., Jones, B., Bergmann, J., Clement, V., Ober, K., ... & Midgley, A. (2019). *Groundswell: Preparing for internal climate migration*. World Bank.

Ritchie, H. (2019). *Who has contributed most to global CO_2 emissions?* Our World in Data. https://ourworldindata.org/contributed-most-global-co2. Accessed 19 May 2022.

Sapiains, R., Ugarte, A. M., Aldunce, P., Marchant, G., Romero, J. A., Gonzalez, M. E., & Inostroza-Lazo, V. (2020). Local perceptions of fires risk and policy implications in the hills of Valparaiso, Chile. *Sustainability, 12*(10).

Salzinger, M., & Desmidt, S. (2021). *Climate change and conflict in the Central Sahel: A shared responsibility to support local resilience*.

Tan, X., Wang, Y., Gu, B., Kong, L., & Zeng, A. (2022). Research on the national climate governance system toward carbon neutrality—A critical literature review. *Fundamental Research, 2*(3), 384–391. https://doi.org/10.1016/j.fmre.2022.03.010. Accessed 19 May 2022.

Taylor, A. (2014, March). *Anton Cartwright and Catherine Sutherland, institutional pathways for local climate adaptation: A comparison of three South African municipalities*. AFD Focales Series.

Termeer, C., Dewulf, A., & Biesbroek, G. (2017). Transformational change: governance interventions for climate change adaptation from a continuous change perspective. *Journal of Environmental Planning and Management, 60*(4), 558–576. https://doi.org/10.1080/09640568.2016.1168288. Accessed 14 May 2022.

Transparency.org. (2022). *Nigeria*. https://www.transparency.org/en/countries/nigeria. Accessed 2 June 2022.

Wambua, C. (2019). The Kenya climate change act 2016: Emerging lessons from a pioneer law. *Carbon & Climate Law Review: CCLR, 13*(4), 257–269.

Wildavsky, A. (1973). If planning is everything, maybe it's nothing. *Policy Sciences, 4*(2), 127–153.

Worker, J. (2017). National climate change governance-Topic guide-August 2016.

CHAPTER 4

The Role of Indigenous and Experiential Knowledge in Advancing Enhanced Climate Action and Policymaking in Africa

Caiphas Brewsters Soyapi, Michael Addaney, and Habib Sani Usman

1 Introduction

The chapter examines the usefulness of indigenous and experiential knowledge (IEK) in climate change action in Africa. Climate change is globally recognised as a leading human and environmental crisis of the twenty-first century and that it is human-induced and demands

C. B. Soyapi (✉)
North-West University, Potchefstroom, South Africa
e-mail: Brews.Soyapi@nwu.ac.za

M. Addaney
Department of Land Management, University of Energy and Natural Resources, Sunyani, Ghana
e-mail: michael.addaney@uenr.edu.gh

Centre for Public Management and Governance, University of Johannesburg, Johannesburg, South Africa

M. Addaney et al. (eds.), *Climate Change in Africa*,
https://doi.org/10.1007/978-3-031-30050-9_4

collective action at all levels (Atapattu, 2016; Blair & Rajendra, 2006; Mayer, 2018). There is a broad consensus among scientists and practitioners from diverse backgrounds that climate change will have devastating, yet differential impacts on different geographic regions unless radical mitigation and adaptation responses are effectively implemented (Addaney & Cobbinah, 2019; Schellnhuber et al., 2006). There are two main responses to climate change namely mitigation and adaptation. The latter is theorised as the process of adjusting to the current and projected effects of climate change (Abbass et al., 2022; IPCC, 2007). Mitigation is conceived as minimising the severity of the impacts of climate change through reducing the emission of greenhouse gases (GHG) into the atmosphere (Raman et al., 2012). Due to Africa's geographical location, per capita contribution to global GHG emissions and level of development (Addaney & Cobbinah, 2019), the revised African Climate Change Strategy adopted in 2021 prioritises adaptation action considering that adaptive responses have potential mitigation benefits. Several studies (e.g. Landauer et al., 2019; Rohe et al., 2017) have observed the complementary and interconnected nature of mitigation and adaptation action. Other studies (e.g. de Oliveira et al., 2016; Gao et al., 2018; Seto et al., 2014) have also stressed the co-benefits of adaptation and mitigation policies and plans. For instance, Landauer et al. (2019) argue that adaptation reduces a region or an areas' sensitivity to climate risks while mitigation lessens their exposure. On the other hand, Jones (2007) showed differing understandings on the interrelatedness of mitigation and adaptation and, thus, proposed that the two concepts should be treated differently, especially considering the diversity of continental scale and policy design. Nonetheless, adaptation and mitigation policies are critical for the continuous development of African countries and communities, and carefully combining them offers enormous benefits, increases synergies and minimises conflicts (Grafakos et al., 2019). As emphasised by Cobbinah and Addaney (2019, 2022), African countries have given attention to both adaptation and mitigation policies in planning.

H. S. Usman
School of Law, American University of Nigeria, Yola, Nigeria
e-mail: habib.usman@aun.edu.ng

At the regional level, the 54 countries on the continent contribute little to greenhouse gas emissions, but will seriously suffer from the undesirable impacts of climate change due to underdevelopment, persistent poverty and weak governance systems. For instance, African countries are collectively responsible for less than 5% of global greenhouse gas (GHG) emissions (African Union, 2020), yet Africa will bear a disproportionate burden of the catastrophic impacts of climate change. For this reason, African countries have a definitive part to play on the global fight against climate change and Africa also needs pragmatic solutions. Understanding climate change, its impacts and appropriate mitigation and adaptation measures are the major challenges confronting African governments and communities. For instance, changes in hydroclimatic variables such as rainfall and temperature in response to greenhouse gas emissions have been detected with a good level of confidence in many parts of the world, but detecting these changes remains uncertain over Africa (Al-Zu'bi et al., 2022). The main reason for this is that Africa has a less well-developed scientific observation network relative to other continents. For instance, the percentage of stations on the continent in line with these standards dropped from 57% in 2011 to 22% in 2019 due mainly to accessibility issues, conflicts and lack of investments (Al-Zu'bi et al., 2022). This problem is worsened by a huge and constant decline in the number of observation stations fully meeting the standards set by the World Meteorological Organization. However, if international and continental climate action is to be enhanced and sustained, there must be innovative and alternative approaches of constructing climate information to inform the development of policies and programmatic initiatives to ensure cost-effectiveness and bottom-up action (Anaafo, 2019; Adusu et al., 2023).

The African continent is characterised by recurrent extreme weather events (e.g. droughts and extreme heat) which have been increasing in intensity and frequency over the past 100 years with marked rainfall declines and droughts that exceed those projected by future climate models (Benson & Clay, 2003; Brooks, 1999; Hulme et al., 2001). The survival of African communities and populations is evidence that they have developed indigenous and experiential mechanisms and strategies to adjust to these extreme weather events like droughts (Nyong et al., 2007). Experiential knowledge on how local populations have coped with previous extreme weather and climate events has the potential of providing important signposts for addressing current and future events (Adusu et al., 2023). To this end, IEK has become increasingly recognised in global climate action. For instance, the 2015 Paris Agreement underscores the significance of incorporating local and indigenous knowledge in enhancing understanding of climate change and developing appropriate actions for adaptation and mitigation (Article 7.5). The UN sustainable development goals (SDGs) also emphasise IEK as a cross-cutting issue which may be deployed in realising the SDGs, especially in addressing food insecurity and poverty (UN Sustainable Development Goals, 2015; Leal Filho et al., 2021). Thus, there is a need to explore alternative ways of developing climate change mitigation and adaptation measures that are relevant to the local context within which they are produced and implemented. However, integrating IEK into climate change policies and strategies should not be done at the expense of modern scientific knowledge but should rather complement global knowledge systems. That is the niche within which this study is situated. The chapter therefore discusses the utility and benefits of IEK's potential in reducing vulnerability to climate change and/or improving the resilience of communities with reflections on barriers that deter the mainstreaming of IEK into climate action and strategies. In achieving this, the chapter uses two main methods: a systematic review of relevant literature and cross-case analysis of case studies from selected African countries where IEK informs climate action and options, including indigenous land use practices and weather prediction focusing on the effectiveness of IEK in reducing vulnerability and the impacts of climate change. The discussion starts by briefly exploring the definitional dimensions and potential value of IEK in climate change debates. Thereafter, the discussions consider some ideas on how IEK could be mainstreamed into climate adaptation action. In the last section, an attempt is made of a case for the need to upscale IEK in climate change strategies and policy processes. As a caveat, however, the discussion does not fall into any particular dimension of IEK (forestry, agricultural, weather patterns, etc.) but rather focuses on the generalities and commonalities from the literature.

2 Understanding Indigenous and Experiential Knowledge: Meaning and Application to Climate Action

Countries and communities in developing regions are the earliest and hardest hit of climate change impacts (UNISDR, 2015). Thus, there is a broad consensus that IEK plays a central role in understanding the local context of climate change impacts and risk (Balay-As et al., 2018). IEK has been conceptualised by Bwambale et al. (2020) as views that are culturally built by indigenous people concerning the reality they experience. The critical role of IEK has also been recognised by the UN Sendai Framework for Disaster Risk Reduction (UNISDR, 2015) by emphasising the suitability of indigenous knowledge in complementing scientific knowledge to understand climate change impacts and risks. Ludwig (2017) argues that IEK exposes blind spots of local regularities that are invisible to modern scientific methodologies in related fields. In climate science and related studies, the significant claim is that by experiencing recurrent extreme climate events, indigenous communities accumulate an indigenous geography of climate change adaptation and mitigation measures. The lived experiences of indigenous local communities acquaint them with specific timings and locations of occurrence, the elements frequently affected and the processes that precede extreme events (Hiwasaki et al., 2014; Tran et al., 2009). They thus accumulate indigenous experiences leading to development of local theories and/or worldviews on climate change impacts, risks and vulnerabilities. Apart from experience and familiarity with the locale, indigenous understanding evolves under the influence of diverse socioeconomic and ecological factors. Thus, IEK is the basis for local environmental decision-making in many rural communities across Africa as it has value for the culture in which it evolves and for scientists and policymakers engaged in sustainable community development.

In December 2021, the Eastern Cape Division of the South African High Court interdicted Shell Exploration and Production South Africa BV (Shell) from continuing with seismic survey activities off the South African eastern Coast. In painting the picture of how some traditional communities were aggrieved by the decision of the authorities to grant an exploration licence to Shell, the court narrated the evidence of some of the applicants as follows:

> Mr Zukulu remembers how, in his earlier days, members of the Jama village (where he was born and which is about 15 kms from the coastline) would walk to the sea to collect its healing waters and sacred sand, harvest mussels, catch fish and would sometimes camp on the coast. His mother's grandfather would go to the ocean every month to cleanse himself for healing purposes. Not only the land, but also the sea, plays an important role in the way of life of members of the Amadiba traditional community. ... Their dependence on and the significance of the sea made training in conservation and sustainable harvesting on land and sea imperative. Mr Zukulu said that they "are the conservationists of the sea in our area, using practices handed down to us over generations" ("*Sustaining the Wild Coast NPC v Minister of Mineral Resources and Energy*", 2022).

This anecdotal evidence is reminiscent of the African stories of old, where such information is then passed on to each generation as a way of preserving tradition and practices. Anecdotal as it might have been in this case, the High Court then added its perception to the fray, noting that, "[a]s indigenous peoples, they feel responsible for conserving the planet for themselves and humanity" (*Sustaining the Wild Coast NPC v Minister of Mineral Resources and Energy*). The message was simple: those whose lives stand to be impacted by development activities have a say in such activities. Part of the complaints around Shell's exploration activities where that there had been no climate considerations, and that ultimately, the project would impact on the environment disproportionately, compared to the potential benefits. The anecdotes, along with mainstream data on exploration activities ultimately resulted in the High Court finding against Shell.

Anecdotes of these kind IEK, otherwise defined as "the intellectual behavior and beliefs of indigenous societies or local information about the relationship of living beings (including humans) with one another and with their environment" (Kupika et al., 2019). In view of the utility of such beliefs or information to the survival of some communities, IEK as a potential source of valuable knowledge for climate adaptation takes on a 'visibilist' (as opposed to the 'invisibilist') outlook (Reyes-García et al. 2016), as this would imply that the effects of climate change can be observed by the naked eye, and that such observations are useful to understanding not only personal experiences, but also localised experiences of changes in climate patterns. IEK involves communal practices as well as social learning. In such communal settings, individuals learn from each other, often through trial and error. This is why experiential learning is also conceptualised as 'learn as you go' and is considered to involve 'iterative, reflective and feedback mechanisms' (Ndebele-Murisa et al., 2020). Others label IEK as 'place-based' climate forecasting, indicating that such IEK is usually localised (Spear et al., 2019). Yet others described such experiences as 'user risk perceptions', focusing on how, for example, those who primarily rely on the environment for their livelihoods, perceive of the risks associated with climate-related changes to their environment (Steynor & Pasquini, 2019). In other words, IEK is significant for its capacity to bring forth contextual nuances to the climate adaptation discourse, which ultimately makes IEK place, time- and culture-specific (Codjoe et al., 2014). This is particularly significant if regard is had to Africa's context of a developing region fraught with poverty, underdevelopment and with an estimated 50 million people reliant on environment-based means of survival (Nyadzi et al., 2021).

When applied directly to the environment, IEK is also labelled 'local ecological knowledge', representing "knowledge, practices, and beliefs shared among local resource users regarding ecological interaction within ecosystems" (Kupika et al., 2019). From such environmental lens, IEK brings to the fore sensitivities, exposure levels and the adaptive capacity of communities. Now considering that from such environmental perspective, climate variability is a concern and reality that traditional or indigenous communities have lived with for time immemorial, there is a case to be made that such communities have been adapting to climate change impacts. For example, there are populations living in semi-arid rural environments who have been finding ways to live with the droughts and unpredictable rainfall patterns (García-del-Amo et al., 2020; Spear et al., 2019). This is what others call indigenous technical science and philosophy because it involves diverse ecological, seasonal and phenological knowledge (Kupika et al., 2019; Petzold et al., 2020). In fact, history informs of how traditional communities were resourceful and practical. For example, they were nomadic and were practical enough to see the need to move to a different area when environmental pressures mounted. Pastoralism and water harvesting are examples of methods that developed after the wars and droughts associated with the Mfecane in the eighteenth century (Kupika et al., 2019; Murombedzi, 2003). Considering that the environment was integral to the cultural and social life of such communities (including medicinal and food sourcing), these were adaptive innovations which had practical and useful utility for the time (Amechi, 2009). Accordingly, for all the vulnerabilities and troubles they meet in the face of climate change, many indigenous communities are not without means or knowledge to adapt and mitigate climate-related challenges.

As such, although scientific and global repositories of climate information provide accurate details of climate change, the catastrophe it is bringing and will bring, such information can potentially fall short of providing nuanced localised and experiential forms of knowledge and observations (García-del-Amo et al., 2020). It is this personalised and experienced knowledge, at times hidden in far remote places, which could be crucial to understanding localised adaptive capacity of different communities. To the extent that it is accepted that climate change impacts are not manifesting in a vacuum, then it stands to reason that those who have experienced some of the challenges associated with climate change have a contribution to make to the adaptation discussion.

3 Mainstreaming Indigenous and Experiential Knowledge in Climate Action in Africa

IEK manifests in uniquely diverse ways. For example, to forecast changes in weather patterns, some communities rely on ethnobotanical and ethnozoological forms of knowledge which can be based on some indicators from astrological, atmospheric and biological phenomena (Kupika et al., 2019). The appearance or blossoming of certain flowers, the appearance of or changes in the behaviour of some wild insects and even the conditions of the clouds or the appearance of the moon are all examples of how IEK is understood (García-del-Amo et al., 2020; Kupika et al., 2019; Spear et al., 2019). These considerations would then, for example, provide guidance as to what crops to plant and when to plant them. While some communities who are involved in the make-up of indigenous and experiential agricultural, fishing or forestry knowledge are aware of changes in weather patterns, they are generally not aware that climate change is the underlying cause (Nyadzi et al., 2021; Selormey et al., 2019). To be sure, in some cases, 4 in every 10 Africans are not familiar with climate change while 3 in 10 are fully climate literate (Selormey et al., 2019). Arguably, the majority of these people that are unaware of climate change or those not fully literate could be the very same people that have been experiencing climate change impacts and adapting to such impacts over time. This is primarily because such communities have experienced severe droughts in increasingly longer periods, with lower agricultural produce as a result of extreme weather patterns such as flooding (and droughts). Tellingly, those involved in agriculture and those residing in rural communities have much to say regarding changing weather patterns, going as far as noting that life is worsening because of climate change (Selormey et al., 2019).

Given how climate variability results in Africa's obvious vulnerability, approaches to climate change adaptation will need to also be contextual and localised. IEK could become valuable because of its potential as a lived reality. Arguably, it is reasonable to assume that a local fisherman who relies on fishing for survival can tell when fish stocks are low, or when they have to delve deeper into waters for fish. A pastoralist could be able to accurately tell the landscape, the grazing patterns and any changes related thereto, and a crop farmer will likely have key knowledge of how produce varies, depending on changing weather patterns. For these reasons, because there is an intimate link between the material conditions of life for these communities and the state of their environment, their experiences are first-hand. A deteriorating agricultural, fishing or forestry space can have both direct and indirect impacts on the quality of life of such communities. Yet, as confirmed in Sustaining the Wild Coast NPC v Minister of Mineral Resources and Energy, (2022) many such communities have historically adapted to the varying impacts of environmental change as many have lived to tell the stories through indigenous and experiential knowledge.

Nevertheless, even as the potential for IEK is palpable, the potential for bias cannot be ignored. Personalised experiences are too often informed by the worst of experiences which could be far removed for long-term changes in climate. For example, there is evidence that a heat spell or other short-term fluctuation in weather patterns can elicit exaggerated claims of changes in climate (Reyes-García et al., 2016). As such, the unusual event (as opposed to the normal or slow change processes) will likely elicit some form of perception on the changes in the weather. In fact, there is a legitimate belief that personal experiences of climatic changes elicit strong emotional responses, especially when related to a sense of meaning that people could have to a place or its environment (Steynor & Pasquini, 2019). This is why IEK cannot, of itself, be the basis of mobilised state-led adaptation strategies. This is particularly true considering that IEK, like customary law, is often unwritten, passed on from generation to generation. Its localised character means that there might be varying accounts to work with if one seeks to construe some common thread running through community practices in climate adaptation. To this end, IEK is unlikely to be value-neutral, with the result that caution must be exercised when IEK is elicited, ensuring that information related to actual long-term changes affecting essential processes such as agriculture, forestry or fishing is combined with scientific knowledge.

Furthermore, one would have to navigate scepticism associated with the accuracy of IEK, owing to the fact that if there is actually no scientific grounding of climate change phenomena, how could descriptions be captured sufficiently enough (without scientific instruments and inquiries) to justify acting upon them (García-del-Amo et al., 2020). Furthermore, in the event that numerical meteorological forecasts are presented to some communities, the existence of IEK could inhibit the use of the forecasts, especially when presented to those with strong religious and traditional beliefs in relation to lands and weather patterns (Spear et al., 2019). The assumption here is that such farmers, for example, would rather use the 'familiar' or the 'experiential' as tried and tested modes of aiding in decision-making. More so, in other instances there are observations to the effect that some farmers believe that rain is from the gods or ancestors, with the result that for them, no human intervention, apart from religious ones, could impact on whether rains fall or not (Nyadzi et al., 2021; Spear et al., 2019). Such religious or traditional value base has the implication of creating various methods or processes of adaptation, owing to the varied religious and traditional belief systems of communities.

These cautionary notes notwithstanding, one thing is true, climate change is consequential. IEK could therefore become a repository for strengthening the resilience of communities who have already been adapting to climatic changes. The combined efforts of mainstream meteorologically driven climate adaptation plans and IEK could offer a more integrated understanding of scientific and local accounts of climate change. Consequently, considering that climate services are often "developed with a superficial understanding of the target audience, are data-driven and are dominated by the western assumptions and cultural norms of those developing them" (Steynor & Pasquini, 2019), we believe that IEK can inform climate science in relation to the practical and often latent socioeconomic effects of climate change, while also giving an identifiable human face to the problems, as opposed to generalised accounts. Local people with indigenous knowledge can thus be allies in the quest for adaptation as their role will unlikely supplant that which science provides (Reyes-García et al., 2016; Nyadzi et al., 2021). Furthermore, considering that adaptation strategies based on IEK are appreciated by small-scale farmers in African and developing countries, it is accepted that the mainstreaming of adaptation initiatives based on IEK into formal climate change adaptation governance in Africa would not present much problem (Ajani et al., 2013).

At this point, a crucial aspect is clarifying the benefits of, as well as specifying how mainstreaming IEK in climate change adaptation action in Africa could occur. Mainstreaming itself has been described as critical to addressing development and climate change problems (Schaar, 2008), yet different terms are used interchangeably in describing mainstreaming as a process. For example, Mickwitz (2009) portrays mainstreaming as the integration of climate change governance policy into development planning, while others describe mainstreaming as the alignment of climate change adaptation strategies with development planning across all sectors of the economy (de Coninck, 2009). Mainstreaming of IEK in climate change adaptation action in Africa is associated with several benefits. For a start, through mainstreaming, institutional conflicts between adaptation strategies based on scientific knowledge and IEK could be drastically minimised (Oulu, 2011). Mainstreaming could also assist in minimising financial loss due to the impacts of climate change, and it could promote the consolidation of resources meant for development activities (Persson & Klein, 2008). In the opinion of Craig (2010), mainstreaming IEK into cross-sectoral development planning could promote the planning and management of cross-sectoral adaptation policies and decision-making process (Craig, 2010). Climate change governance institutions in Africa are therefore at an advantage of mainstreaming IEK into agricultural development plans that are aimed at developing the resilience of crop varieties, managing lands and forests, as well as boosting water availability (De Witt, 2018). Given that African countries are already strained by lack of good governance, poor technology and endemic poverty which make it difficult to adopt and solely rely on adaptation initiatives based on modern science, IEK could be used to achieve localised adaptation results. This would be particularly beneficial for small-scale farmers who generally find adaptation initiatives based on modern scientific knowledge to be sophisticated and complicated (Chirisa et al., 2018). This is precisely why, for example, small-scale farmers in Southern Africa mostly adopt IEK as a suitable and less complex method of promoting subsistence crops production (Mugambiwa, 2018).

Consequently, we believe that IEK could be mainstreamed into climate adaptation action through processes of integration through horizontal and vertical governance efforts. Mainstreaming takes a horizontal dimension when it involves the integration of adaptation measures by different sectors of the economy or different departments of the same government (Lafferty & Hovden, 2003). This type of mainstreaming could minimise rift among government ministries, departments and agencies (Lafferty & Hovden, 2003). A vertical dimension to mainstreaming involves the integration of adaptation into development concerns by different levels of the same government (Lasco et al., 2009). An illustration of vertical dimension to mainstreaming is seen where for instance, a department for water resources at local government level incorporates strategies for irrigating drylands as provided by the ministry for agriculture at the national level (Mickwitz et al., 2009).

Consequently, both vertical and horizontal forms of mainstreaming of IEK could be applied in strengthening local climate change adaptation institutions in Africa (Mararike, 1999). Local climate change institutions could perform optimally in driving adaptation initiatives when guided by existing IEK, than when they are compelled to function based on complex and modern scientific knowledge. For example, adaptation strategies such as the selective keeping of livestock and the planting of early maturing crops in areas with less rainfall, could be implemented by local people who require little or no training in a participatory, cost-effective and sustainable manner (Mutambisi et al., 2020). Thus far, we know that any environmental conservation efforts will not be successful without communities, particularly indigenous ones who will also disproportionately experience the impacts of climate change for example. This is precisely why the South African Supreme Court of Appeal, in a 2018 conservation case ("Gongqose case", 2018) observed that,

> it is important to remember that as regards conservation and long-term sustainable utilization of marine resources in the MPA [Marine Protected Area], the Dwesa-Cwebe communities have a greater interest in marine resources associated with their traditions and customs, than any other people. These customs recognize the need to sustain the resources that the sea provides.

Ultimately, customs have no doubt served traditional communities well in the past. These processes, however, remain outside mainstream science. Yet, when observed, recorded and reviewed, they could have an impact on the choices and options for adaptation beyond specific localities. Therefore, IEK could benefit from a more positivistic stance. The benefit is that the knowledge holders exist and the communities are available. If concerted efforts are made to record how these communities have been adapting to changes in climate, then such knowledge becomes accessible. To this end, a functional avenue would be to pursue targeted approaches for sectoral information. To understand experiences of changes in agriculture, information should be sought from agriculturalists. The same goes for pastoralists, etc. Once targeted, there could be the establishment of centralised information repository centres where information can be relayed on a 'place-based' process of localised data capturing of community-based adaptation strategies.

4 Conclusion

This chapter has provided an overview of the role and extent to which IEK could support climate change policies and strategies. The discussion shows that, in different contexts, this knowledge can open pathways for using prior experience as a tool to address particular climate events. For instance, the recording of IEK presupposes the involvement of local knowledge holders in the process. This is crucial for strengthening the legitimacy of climate change interventions in local communities since IEK holders "can point to potential organisms, ecosystems and locations for future research focus, provide alternative lines of evidence for data or models…, and be used as illustrations or case examples in communicating potential consequences of climate change to broader audience" (Prober et al., 2019). Accordingly, IEK holders deserve a seat at the table, especially if decisions from such table have a bearing on how climate action is to be pursued. Ultimately, the inquiries should be informed by the imperative for inclusive and enhanced climate action, such that IEK works alongside mainstream scientific knowledge in a complimentary, mutually beneficial and locally (for communities) understandable manner. What remains true is that regardless of the currency afforded to IEK in climate change policies, a diversity of adaptation and mitigation strategies, processes or approaches will always bode well for communities. Two major problems that can be identified as obstacles to integrating IEK into official climate change strategies include the recognition of the need to, and how to actually assimilate IEK into modern science. Furthermore, experiential knowledge is increasingly exhibiting a similitude with scientific methods as many ideas in indigenous knowledge that were once seen as primitive and misguided are now understood as appropriate and urbane. African governments, scholars and practitioners should therefore deliberately recognise the utility of IEK in offering local communities with the capability of contributing to climate action. Finally, there is a pressing need to adopt bottom-up participatory approaches in the design and implementation of climate change interventions to boost the highest level of local participation.

References

Abbass, K., Qasim, M. Z., Song, H., Murshed, M., Mahmood, H., & Younis, I. (2022). A review of the global climate change impacts, adaptation, and sustainable mitigation measures. *Environmental Science and Pollution Research, 29*(28), 42539–42559.

Addaney, M., & Cobbinah P. B. (2019). *Climate change, urban planning and sustainable development in Africa: The difference worth appreciating*. The Geography of Climate Change Adaptation in Urban Africa (pp. 3–26).

Adusu, D., David A., Simon A., & Michael A. 2023. Experiential Knowledge of urbanites on climatic changes in the Sunyani municipality, Ghana. *Journal of Urban Affairs, 45* (3), 488–504.

Ajani, E. N., Mgbenka, R. N., & Okeke, M. N. (2013). *Use of indigenous knowledge as a strategy for climate change adaptation among farmers in sub-Saharan Africa: Implications for policy.*

Al-Zu'bi, M., Dejene, S. W., Hounkpè, J., Kupika, O. L., Lwasa, S., Mbenge, M., ... & Touré, N. D. E. (2022). African perspectives on climate change research. *Nature Climate Change*, 1–7.

Amechi, E. P. (2009). Enhancing environmental protection and socio-economic development in Africa: A fresh look at the right to a general satisfactory environment under the African charter on human and peoples' rights. *Law Environment and Development Journal, 5*, 58.

Anaafo, D. (2019). Between science and local knowledge: Improving the communication of climate change to rural agriculturists in the Bolgatanga Municipality, Ghana. *Local Environment, 24*(3), 201–215.

Atapattu, S. (2016). Climate Change, International Environmental Law Principles, and the North-South Divide. *Transnat'l L. & Contemp. Probs, 26*, 247.

Balay-As, M., Marlowe, J. & Gaillard, J. C. (2018). Deconstructing the binary between indigenous and scientific knowledge in disaster risk reduction: Approaches to high impact weather hazards. *International Journal of Disaster Risk Reduction, 30*, 18–24.

Benson, C., & Clay E (2003). *Economic and financial impacts of natural disasters: An assessment of their effects and options for mitigation: Synthesis report.* Overseas Development Institute.

Blair, T., & Rajendra, P. (2006). Avoiding dangerous climate change. Cambridge University Press, 17.

Bwambale, J. (2020). *The impact of climate change on the interactions between water, energy and food systems in the Victoria Nile Sub-Basin.* Master's thesis, PAUWES.

Chirisa, I., Matamanda, A., & Mutambwa, J. (2018). Africa's dilemmas in climate change communication: Universalistic science versus indigenous technical knowledge. *Handbook of climate change communication* (Vol. 1). Springer.

Codjoe, S. N., Ardey, G. O., & Burkett, V. (2014). Perception, experience, and indigenous knowledge of climate change and variability: The case of Accra, a sub-Saharan African city. *Regional Environmental Change, 14*, 369–383.

Craig, R. K. (2010). Stationarity is dead-long live transformation: Five principles for climate change adaptation law. *Harvard Environmental Law Review, 34*, 9.

de Coninck, S. (2009). *Mainstreaming poverty-environment linkages into development planning: A handbook for practitioners.* UNEP/Earthprint.

de Oliveira, Jose A. P., & Christopher N. H. D. (2016). Governance and networks for health co-benefits of climate change mitigation: Lessons from two Indian cities. *Environment International, 97*, 146–154.

De Witt, S. (2018). Measuring our investment in the future. *African Evaluation Journal, 6*, 1–3.

Gao, C., Kalev, K., Per-Olof Ö., & Tord K. (2018). Occupational heat stress assessment and protective strategies in the context of climate change. *International Journal of Biometeorology, 62*, 359–371.

García-del-Amo, D., Graham Mortyn, P., & Reyes-García, V. (2020). Including indigenous and local knowledge in climate research: An assessment of the opinion of Spanish climate change researchers. *Climatic Change, 160*, 67–88.

"Gongqose v Minister of Agriculture, Forestry and Fisheries." In 2018. (5) SA 104 (SCA).

Grafakos, S., Trigg, K., Landauer, M., Chelleri, L., & Dhakal, S. (2019). Analytical framework to evaluate the level of integration of climate adaptation and mitigation in cities. *Climatic Change, 154*, 87–106.

Hiwasaki, L., Emmanuel L., & Rajib S. (2014). Process for integrating local and indigenous knowledge with science for hydro-meteorological disaster risk reduction and climate change adaptation in coastal and small island communities. *International Journal of Disaster Risk Reduction*, 10, 15–27.

IPCC fourth assessment report. (2007). *The physical science basis, 2*, 580–595.

Jones, G. V. (2007). Climate change: Observations, projections, and general implications for viticulture and wine production. *Economics Department-Working Paper*, *7*, 14.

Kupika, O. L., Gandiwa, E., Nhamo, G., & Kativu, S. (2019). Local ecological knowledge on climate change and ecosystem-based adaptation strategies promote resilience in the Middle Zambezi Biosphere Reserve, Zimbabwe. *Scientifica*.

Lafferty, W., & Hovden, E. (2003). Environmental policy integration: Towards an analytical framework. *Environmental Politics*, *12*, 1–22.

Landauer, M., Juhola, S., & Klein, J. (2019). The role of scale in integrating climate change adaptation and mitigation in cities. *Journal of Environmental Planning and Management*, *62*(5), 741–765.

Lasco, R. D., Pulhin, F. B., Jaranilla-Sanchez, P. A., Rafaela Jane, P., Delfino, R. G., & Garcia, K. (2009). Mainstreaming adaptation in developing countries: The case of the Philippines. *Climate and Development*, *1*, 130–146.

Leal F., Walter, V. O. L., Markus W., Amanda L. S., & Fernanda F. (2021). Poverty: A central barrier to the implementation of the UN Sustainable Development Goals. *Environmental Science & Policy*, *125*, 96–104.

Ludwig, D. (2017). The Objectivity of Local Knowledge: Lessons from Ethnobiology. *Synthese*, *194*12, 4705–4720.

Mararike, C. G. (1999). *Survival strategies in rural Zimbabwe: The role of assets, indigenous knowledge, and organisations*. Mond Books.

Mayer, B. (2018). Obligations of conduct in the international law on climate change: A defence. Review of European. *Comparative & International Environmental Law*, *27*(2), 130–140.

Mickwitz, P., Aix, F., Beck, S., Carss, D., Ferrand, N., Görg, C., Jensen, A., Kivimaa, P., Kuhlicke, C., & Kuindersma, W. (2009). *Climate policy integration, coherence and governance* (irstea).

Mugambiwa, S. S. (2018). Adaptation measures to sustain indigenous practices and the use of indigenous knowledge systems to adapt to climate change in Mutoko rural district of Zimbabwe. *Jàmbá: Journal of Disaster Risk Studies*, *10*, 1–9.

Murombedzi, J. (2003). Pre-colonial and colonial conservation practices in Southern Africa and their legacy today. *Gland, Switzerland: IUCN*.

Mutambisi, T., Chanza, N., Matamanda, A. R., Ncube, R., & Chirisa, I. (2020). Climate change adaptation in Southern Africa: Universalistic science or indigenous knowledge or hybrid. In *African handbook of climate change adaptation*.

Ndebele-Murisa, M. R., Mubaya, C. P., Pretorius, L., Mamombe, R., Iipinge, K., Nchito, W., Mfune, J. K., Siame, G., & Mwalukanga, B. (2020). City to city learning and knowledge exchange for climate resilience in southern Africa. *PLoS ONE*, *15*, e0227915.

Nyadzi, E., Ajayi, O. C., & Ludwig, F. (2021). *Indigenous knowledge and climate change adaptation in Africa: A systematic.*

Oulu, M. O. (2011). Mainstreaming climate adaptation in Kenya. *Climate Law, 2*, 375–394.

Persson, A., & Klein, R. J. T. (2008). *Mainstreaming adaptation to climate change into official development assistance: Integration of long-term climate concerns and short-term development needs*. Stockholm Environment Institute.

Petzold, J., Andrews, N., Ford, J. D., Hedemann, C., & Postigo, J. C. (2020). Indigenous knowledge on climate change adaptation: A global evidence map of academic literature. *Environmental Research Letters, 15*, 113007.

Prober, S. M., Raisbeck-Brown, N., Porter, N. B., Williams, K. J., Leviston, Z., & Dickson, F. (2019). Recent climate-driven ecological change across a continent as perceived through local ecological knowledge. *PLoS ONE, 14*, e0224625.

Raman, A., Verulkar, S. B., Mandal, N. P., Variar, M., Shukla, V. D., Dwivedi, J. L., et al. 2012. Drought yield index to select high yielding rice lines under different drought stress severities. *Rice*, *5*, 1–12.

Reyes-García, V., Fernández-Llamazares, Á., Guèze, M., Garcés, A., Mallo, M., Vila-Gómez, M., & Vilaseca, M. (2016). Local indicators of climate change: The potential contribution of local knowledge to climate research. *Wiley Interdisciplinary Reviews: Climate Change, 7*, 109–124.

Rohe, J. R., S. Aswani, A. Schlüter & S. C. Ferse, (2017). Multiple Drivers of Local (Non-) Compliance in Community45 Based Marine Resource Management: Case Studies from the South Pacific. *Frontiers in Marine Science*, *4*, 172.

Schaar, J. (2008). Overview of adaptation mainstreaming initiatives. In *Commission on Climate Change and Development*, 1.

Schellnhuber, H. J., Wolfgang C., Nebojsa., Tom W., & Gary, Y. (2006). Avoiding dangerous climate change. *Ecoscience*, *14*(1), 134–135.

Selormey, E., Dome, M. Z., Osse, L., & Logan, C. (2019). *Change ahead: Experience and awareness of climate change in Africa.*

Seto, K. C., Dhakal, S., Bigio, A., Blanco, H., Delgado, G. C., Dewar, D., Huang, L., Inaba, A., Kansal, A., Lwasa, S. amp; McMahon, J. (2014). *Climate change 2014: Mitigation of climate change. Contribution of working group III to the fifth assessment report of the intergovernmental panel on climate change*. The World Urbanization Prospects. 923–1000.

Spear, D., Selato, J. C., Mosime, B., & Nyamwanza, A. M. (2019). Harnessing diverse knowledge and belief systems to adapt to climate change in semi-arid rural Africa. *Climate Services, 14*, 31–36.

Steynor, A., & Pasquini, L. (2019). Informing climate services in Africa through climate change risk perceptions. *Climate Services, 15*, 100112.

"Sustaining the Wild Coast NPC v Minister of Mineral Resources and Energy." In 2022. JDR 0033 (ECG).

Tran, P., Rajib S., Guillaume C., & John, N. (2009). GIS and local knowledge in disaster management: a case study of flood risk mapping in Viet Nam. *Disasters*, *33*(1), 152–169.

UNISDR. (2015). Sendai Framework for Disaster Risk Reduction 2015-2030 *United Nations Office for Disaster Risk Reduction (UNISDR)*. (p. 32).

CHAPTER 5

Climate Change, REDD+, and Gendered Benefit Sharing in Forest-Dependent Communities of Africa

Rachel Yeboah Nketiah, Naomi Kumi, and Zenebe Mekonnen

1 Introduction

Climate change is frequently referred to as our generation's greatest economic and environmental problem. Existing studies show that about 51% of Africans believe that climate change is happening and is caused by human activity (González & Sánchez, 2022). There is a broad and

R. Y. Nketiah (✉)
Centre for Climate Change and Gender Studies, University for Energy and Natural Resources (UENR), Sunyani, Ghana
e-mail: rachel.yeboah@uenr.edu.gh

N. Kumi
Department of Atmospheric and Climate Science, University of Energy and Natural Resources (UENR), Sunyani, Ghana

Z. Mekonnen
Ethiopian Environment and Forest Research Institute, Addis Ababa, Ethiopia

M. Addaney et al. (eds.), *Climate Change in Africa*,
https://doi.org/10.1007/978-3-031-30050-9_5

long-held scientific consensus that carbon emissions generated by humans are affecting the earth's climate with potentially devastating consequences (Cook et al., 2016). The impacts of climate change have caused the vulnerability of humans, as well as animal and plant species in many parts of the world, especially Africa. The impact is high on very poor and marginalized communities including women and youths (Table 1). Scientists have been delivering warnings about this problem for several decades, well before the predicted increases in global temperatures started to emerge (Bonneuil et al., 2021). Climate change is an international issue; however, its impact varies depending on the geography and locality, especially in terms of effects, exposure, and adaptation (Addaney & Cobbinah, 2019).

Forest resources are significant because they are at the forefront of mitigating human-caused global warming. The Congo Basin forests (next only to the Amazon), for instance, are critical for trapping CO_2 from the atmosphere. Forest resources must be preserved to help reduce the impact of global climate change since these forests serve as carbon sinks (absorbs carbon dioxide from the atmosphere), and this requires financial allocation to appropriately value and promote their preservation.

REDD+ (Reducing Emissions from Deforestation and Forest Degradation [REDD+])[1] is a primary mechanism for addressing global forest governance, and a climate change mitigation approach is created to encourage impoverished nations to reduce carbon emissions, improve biodiversity, and safeguard indigenous people's rights among others. This notwithstanding, REDD+ also comes with various potential threats that affect the forest-dependent communities. Many studies show that currently, REDD+ projects do not adequately incorporate forest-dependent communities, especially women in their policies. REDD+'s current and potential consequences on communities frequently disrupt local people's livelihoods, approaches, organizations, and socio-cultural systems in a variety of ways, including unequal benefit sharing, food insecurity, the release of new powerful stakeholders, illegal land acquisition, unfair free prior and informed consent, and the introduction of monoculture plantations.

[1] REDD+ is a framework created by the UNFCCC Conference of the Parties (COP) to guide activities in the forest sector that reduces emissions from deforestation and forest degradation, as well as the sustainable management of forests and the conservation and enhancement of forest carbon stocks in developing countries.

Table 1 Key reasons why women are more impacted by climate change in Africa

Reasons for women's vulnerability to climate change impacts	*References*
1. **Lack of control over productive resources**: Access to land provides an important source of resilience for women who may lack other options to adapt to the effects of climate change. Indeed, as of 2010, only 15% of land in sub-Saharan Africa is managed by women. For example, in Ghana and Uganda, male farmers are more likely than female farmers to own agricultural tools and livestock	FAO (2011), Doss et al. (2012), Sellers (2016), McOmber (2020), Caroli et al. (2022), and Mekonnen (2022)
2. **Lack of climate information**: The opportunities for incorporating climate knowledge into climate-sensitive sectors, which could greatly enhance their effectiveness, are largely being missed in Africa. **E.g., Men in Senegal and Uganda tend to have access to better weather information than women that can be used to modify production practices**	Tadesse (2010), Twyman et al. (2014), Diouf et al. (2019), and Henriksson et al. (2020)
3. **Lower rate to adopt new technologies**: The adoption of new technologies is a means to tackle climate change impacts. However, women are less likely to adopt than men • E.g., In Uganda, female-headed households and wives of male farmers are significantly less likely to adopt drought-tolerant maize than male farmers • In Ethiopia, male-headed households are significantly more likely than female-headed households to adopt new crop varieties in response to weather changes	Deressa et al. (2009), Gebrehiwot and van der Veen (2013), Fisher & Carr (2015), and Autio et al. (2021)
4. **Dependency on natural resources**: Women are more dependent on natural resources (forests and water) than men. When these resources are lost or become scarce due to the impacts of climate change, women are forced to suffer in the collection of timber and non-timber forest products as well as water	Babugura (2010), Kalaba et al. (2013), Woittiez et al. (2013), Balama et al. (2016), Becerra et al. (2016), and Stevenson et al. (2016)

(continued)

Table 1 (continued)

Reasons for women's vulnerability to climate change impacts	*References*
5. **Less mobility**: Women are stayed-at home caring for their children and doing household activities. However, men are more mobile during extreme events than women do	Dillon et al. (2011), Gray & Mueller (2012), Abebe (2014), Lama et al. (2021), and Mekonnen (2022)
6. **Fewer meals (food insecurity)**: Women, especially in rural areas, are found to be more likely in reducing the number of meals they eat as an adaptation strategy in response to climate shocks. They even eat leftover foods after feeding the whole family and have fewer options to feed themselves in restaurants and hotels as they are dependent on their husbands for income sources	Kakota et al. (2011), Zakari et al. (2014), FAO (2015), Fuhrman et al. (2020), and Mekonnen (2022)

At the international level, REDD+ projects are overseen by various political entities, with varying levels of negotiation and execution. African governments must incorporate gender into forest management at the management level and advocate for gender equality in environmental policy. Do REDD+ policies help to fight for gender equity in Africa's forest-dependent communities, or do they contradict one another when it comes to tackling the impact of climate change? Can they be controlled and applied to enhance Africa's forest governance? The chapter provides answers focusing on the concepts of non-discrimination and equality as well as participation, benefit sharing, and inclusion in discussing the connection between climate change, REDD+, and gender in reducing vulnerabilities of forest fringe communities in Africa. Here it explores the concept of effective participation and benefits sharing in REDD+ agreements, focusing particularly on the policy frameworks capturing how climate change, REDD+, and gender interplay, especially concerning the rights of women in these policies in reducing vulnerabilities among forest fringe communities in Africa (Ghana, Congo, Mozambique, and Tanzania). This chapter follows a structured and standard process by filling the knowledge gap on climate change, REDD+, and gender justice in forest-dependent communities of Africa through a literature review by examining REDD+ policy in action and practice frameworks in Africa and its climate sensitivity policies to gender and women's rights in practice.

2 Research Approach

Climate change, REDD+, and gender justice in forest-dependent communities in Africa were explored using thematic content analysis. Content analysis is a systematic procedure that allows researchers to analyze textual data to identify and aggregate framings, patterns, constructs, and expressions to address a research problem or question (Krippendorff, 2018). It is also a non-intrusive, non-reactive research strategy that takes advantage of documents' inherent availability, stability, precision, and breadth of coverage (Bowen, 2009). The interpretation is based on extracts and quotes, and the analysis follows a methodological premise. While we are aware of various criticisms of this approach, (e.g., poor retrievability, and skewed selectivity), emerging research shows that content analysis is important in articulating, theorizing, and integrating clear gender action plans into topics in developing gender discourse discussions. Furthermore, this form of study has lately gained popularity as a valuable method for assessing gender equality in a range of diverse but related domains across numerous disciplines (Schmidt & Graversen, 2020).

2.1 Data Collection Methods

We have adopted a systematic review approach including both quantitative and qualitative data from published scientific peer-reviewed literature. A systematic literature search was conducted in January 2022 using the online search engine 'Web of Science', 'Science Direct', 'Google Scholar', 'EBSCO Discovery Service', and 'Environmental Index' (Table 2). James et al. (2016) observed that systematic mapping collates, describes, and catalogues available data directly related to a question of interest. The search term used included REDD+, climate change, forest governance, gender justice, forest-dependent communities of Africa, and so forth with reflections on the different processes of climate change and forest governance associated with numerous challenges, experiences, and prospects that are known to be relevant in Africa (e.g., Ghana, The Congo, Tanzania, and Mozambique). The final search with the iteratively optimized search term yielded 350 papers. Out of the 350 papers, 135 papers met the eligibility test upon examining the relevance of the abstracts to our work.

The publications generated were assessed for eligibility and systematically filtered using the following procedure: a preliminary screening

Table 2 Studies included in the analysis

Databases	*No of publications generated*	*No of publications selected*
Science direct	90	29
Web of science	60	31
Google scholar	110	42
EBSCO/Environmental index	90	33
Total	350	135

Source Author's construct

of publication titles, keywords, and abstracts, followed by a more thorough examination of full-text articles, with the elimination of duplicates. To support the study's applicability in the African setting, a considerable number of published papers by researchers and academics were examined. The study's scope was restricted by focusing on a systematic review research design on climate change, REDD+, and gender justice in African forest-dependent communities, as well as key international and regional policies on climate change and forest governance. It incorporated women's interests, concerns, and rights in promoting inclusivity and transparency among forest fringe communities or forest dwellers as a method for reducing climate change and accomplishing the REDD+ goal.

3 Conceptual Review of Climate Change, REDD+, and Gender Justice

Women are disempowered in the agricultural empowerment index in rural communities in Africa (e.g., Mekonnen, 2022) and there is no exception in forest management including REDD+ programs. Despite its primary function as a mechanism for reducing carbon emissions from forests (climate action SDG13), REDD+ increasingly has to deal with concerns regarding social benefits, well-being, and gender as parts of its mandate (e.g., Paris Agreement, SDG5 gender equality, and women's empowerment) (Gizachew et al., 2017; Larson et al., 2018). Regardless of these facts, REDD+ policies and projects are found to be deficient in bringing women on board in the decision-making process and planning (Brown, 2011, Setyowati, 2012, UNDP, 2016, UN WOMEN, 2021).

The impacts of climate change are enormous and almost everyone is affected globally, but the adverse effects are not equally felt. It is documented that women and children are more vulnerable and often bear the impacts of climate change than men (Addaney & Moyo, 2018; Nagel, 2015). For example, in the recovery process from climate-induced disasters, women are more vulnerable in accessing resources and coping with adaptation strategies. Generally, women are susceptible to poverty (especially in Africa) as compared to their male counterparts because of unequal access to financial and other economic resources, and their little involvement in decision-making in society. Women are also said to experience the adverse effects of climate change differently as they are more reliant on climate-sensitive sectors such as agriculture (Addaney & Moyo, 2018; Skinner, 2011). Extreme climate events like floods, droughts, and heat waves have distinctive effects on gender in Africa and women are the most affected, hence the need to recognize gender-just climate solutions according to UNDP (2011). The adverse effects of climate change may also exacerbate water scarcity and food insecurity leading to increases in food costs. Consequently, there will be resource competition and women are bound to be the more vulnerable ones (Addaney, 2018). For instance, in many rural communities across Africa, it is common knowledge that women and girls are responsible for gathering firewood and fetching water for domestic use. In the Sahel region and during drier seasons in most parts of Africa, they have to walk long distances to fetch water, firewood, and food. This unduly increases their workload since they have to trek even longer distances to search for water, reducing their time for other activities such as education, cascading into the perpetuation of unequal gender relations. This reinforces existing gender stereotypes and puts women at a greater risk of the adverse effects of climate change (Addaney, 2018).

In the last three decades, international, regional, and national frameworks have mushroomed to bring gender issues on board (e.g., The Cancun Agreement of the UNFCCC, 2010). However, many attempts to incorporate gender equality in climate change responses have been confined to simplistic, specific, and short-term technical interventions which have failed to dismantle unequal power structures or to exert a structural impact on closing gender gaps when applied. To this end, gender inequalities, manifested in different areas and sectors, hinder sustainable development, and more importantly affect women's climate change adaptation (Caroli et al., 2022; Revelo, 2021; World Bank,

2021). Figure 1 highlights that many gender and intersecting issues that heighten social vulnerability are further exacerbated by climate change impacts, with dire implications for the role of REDD+ in climate change adaptation. Prevailing gender gaps in information, knowledge, and decision-making on REDD+ programs are likely to affect the distribution of future benefits and burdens from REDD+ (Larson et al., 2016). Regardless of prominent gender gaps and challenges, some opportunities and solutions can support gender mainstreaming in REDD+ processes and programs (Table 3).

Wornell et al. (2015) introduced three levels of gender mainstreaming in REDD+ policies: non-substantive, superficial, and integrated inclusion. Integrated inclusion, as stated by Wornell et al. (2015) is the level at which gender can be mainstreamed into REDD+ to achieve gender justice. This is the point where both men and women are welcomed onto the negotiation desk. This allows discussion on the environment and climate change connected to gender equity because the panel is gender-balanced. This facilitates equal participation on the side of men as well as women. It is the responsibility of African countries to ensure there exists gender equity that satisfies gender justice, that adopts clear political commitment, stakeholders' involvement, gender action policies, and decision-making that is inclusive to be at the forefront. To bring this into reality, African countries will have to mainstream gender into environmental policies in general and into REDD+ policies (Bee, 2017). Gender justice is a means of reducing vulnerabilities as a result of climate change.

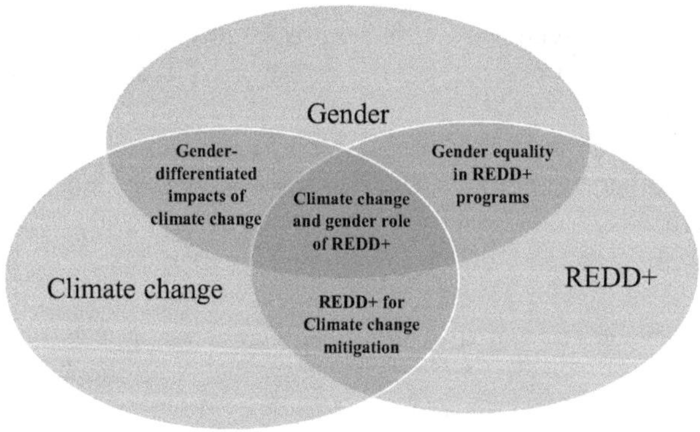

Fig. 1 Visual representation of the gender-climate change-REDD+ nexus

Table 3 Key challenges for women's participation in REDD+ decision-making, solutions, identified best practices and methods of gender mainstreaming in REDD+

Women's key challenges in REDD+ decision-making	*Solutions to the challenges*	*Best practices*
Cultural barriers preventing women from participating in decision-making	• Early interventions: sensitize and build skills of young girls, talk about gender, develop an understanding of climate change • Sensitize men and women together for joint decision-making • Sensitize traditional leaders in the community • Develop different platforms to address unique cultural barriers • Develop programs that encourage behavioral change • Consider the impact of religion and culture and mainstream REDD+ education into religious fora	• Mobilization of community elders and cultural leaders, including women, to promote change • Establishment of women's platforms • Working through local organizations which understand the local culture • Increasing literacy among women • Making use of women-only meetings where appropriate
Insufficient capacity and information to participate in decision-making	• Provide training, skills, and tools to women • Strengthen local non-government organizations (NGOs) by forming national and international networks, using information sharing fora, WhatsApp, and email lists to share data and findings • Identify and support community champions and provide them with skills, public speaking training, and funds to educate others • Promote women's role models and mentors	• Use of existing structures to increase capacity to ensure sustainability • Collaboration with local NGOs • Conducting community sensitization campaigns using local languages • Coordination of REDD+ capacity building efforts with the government to ensure continued support • Conducting exchange visits to best practice projects • Delegation of MPIDO as a learning and a lobbying platform • Packaging information in user-friendly ways: pictures, songs, dance, and radio talk shows • Understanding the needs of different groups of women: leaders, entrepreneurs, young and old and rural

(continued)

Table 3 (continued)

Women's key challenges in REDD+ decision-making	*Solutions to the challenges*	*Best practices*
Insufficient political will and government support for including women in REDD+ decision-making processes	• Mobilize women's platforms and entities with similar motives to lobby the government • Support women candidates to run for office • Engage the media from the beginning • Advocate policymakers with data that show how women are discriminated against and the benefits of engaging women	• Promotion of policy reforms: community land rights, indigenous land or individual ownership • Mobilization of resources from government agencies • Allocation of funds for gender mainstreaming in REDD+
Lack of assets, resulting in food insecurity, and lack of land rights	• Promote alternative income-generating opportunities for women • Promote women-only cooperatives • Promote women-only REDD+ projects • Revise legal frameworks through advocacy: lobby for women's property rights • Follow up on good laws that are not being implemented • Link women's property rights to entrepreneurship training	• Promoting forest use rules that allow women to make decisions on tree-cutting • Introducing new energy technologies to increase women's availability, such as solar and wind • Enhancement of entrepreneurial opportunities and branding of cultural heritage products made by women • Promoting cash crops for women • Developing women trade platforms among African countries
Insufficient donor support for gender issues	• Lobby for further financial support: scale up gender-focused projects that work • Projects should be gender-sensitive from the beginning • Create a fund for women at the national level	• Acknowledgment that each country has its own specific needs, and these are reflected in national strategies • Development of gender strategies on REDD+ and climate change with clear gender outcomes, including for IPs women
Gender mainstreaming methods and tools in REDD+		

Women's key challenges in REDD+ decision-making	*Solutions to the challenges*	*Best practices*
Methods and tools for women networks to participating and engaging in negotiations and advocating for favorable policies and laws in REDD+ in their respective countries include: • Promoting and strengthening national- and regional-level networking • Developing capacity building programs for women networks, CSOs, and IPOs • Promoting access to legal aid, education and outreach for women and local communities on REDD+ • Enhancing lobbying, advocacy, and policy influencing • Promoting media campaigns and awareness programs • Establishing gender equality and climate change clubs in schools • Developing alternative and sustainable livelihoods and social enterprise programs • Embracing new technologies and innovative programs in ERPs and REDD+ • Strengthening research and knowledge management		

Source Adapted from World Bank (2019)

Gender justice comes to play when considering climate change effects on human vulnerability. In the absence of a gender perspective in REDD+, gender justice cannot be achieved.

Benefit sharing is one of the key components of REDD+. Benefit sharing is gaining traction as a governance method to enable more fair and sustainable interactions and results in the face of escalating demand for forests (Soliev et al., 2021). Benefit sharing in this context is achieving an appropriate and long-term benefit allocation to all involved parties. The assumption is that this brings about fairness, inclusiveness, and transparency. This approach also shares the outcomes of the benefit accruing to a particular forest area among the fringe communities or forest dwellers. Scholars increasingly debate benefit sharing as a potential governance approach to transforming existing and potential conflicts in such multi-actor and multi-scale social-ecological systems into more equitable and sustainable interactions and outcomes (Nkhata et al., 2012; Susskind & Ali, 2014; Soliev & Theesfeld, 2017). In forest governance discourse, benefit sharing received particular attention in the last fifteen years under the banner of REDD+ adopted by the UNFCCC (Brockhaus et al., 2014). To achieve appropriate long-term benefit sharing, a deeper knowledge of the articulation of the parties involved and improved forms of benefit sharing, as well as specific ways of arranging them, might help to make REDD+ more successful, efficient, and equitable in Africa. Benefit sharing is a ground for fair, equitable, and collaborative discussion for African counties endowed with natural resources like forests.

3.1 REDD+ Projects in Africa: Challenges, Experiences, and Prospects

REDD+ is a worldwide initiative that uses performance-based incentives to reduce deforestation and forest degradation. REDD+ programs are designed for forest areas where deforestation or degradation is an issue or where there is an opportunity to increase carbon stocks. Assessing the pace of deforestation (or carbon sequestration potential) is indeed a crucial component in the development of a REDD+ project. To properly understand the agents and causes of forest decline, reducing CO_2 emissions from deforestation or forest degradation is necessary. To better understand how REDD+ has been conceived and implemented in the African context in terms of translating leadership commitment to gender inclusion, as well as reflection on success and difficulties, four African countries (Ghana, Congo, Mozambique, and Tanzania) are explored.

Ghana became a REDD+ partner for the first time in 2008. The premise for Ghana building a credible REDD+ project is determining the rate of deforestation and identifying drivers of deforestation and potential concerns. Ghana's REDD+ strategy has been designed to directly meet the requirements of the Warsaw Framework on REDD+ and other COP decisions (Johnson, 2021). The Ghana Cocoa Forest REDD+ Programme (GCFRP) is the first program to be developed under REDD+ in Ghana. It is jointly coordinated by the National REDD+ Secretariat (NRS) of the Forestry Commission (FC), and Ghana's Cocoa Board (Cocobod) (Forestry Commission, 2020).

Dumenu et al. (2014) established that in the country, four different types of benefit sharing arrangements within the forest sector have been identified to achieve viable REDD+ projects. These include: Constitutional Timber Revenue benefit sharing, Modified Taungya System benefit sharing, Commercial Plantation benefit sharing, and Community Resource Management Area (CREMA) benefit sharing. As Ghana moves forward with the implementation of its REDD+ strategy, steps such as identifying a set of activities that will reduce emissions from ongoing deforestation or degradation, as well as a review of existing benefit sharing mechanisms, were considered for critical assessment under these four benefits sharing categories. The topic of discussion is that the REDD+ initiative in Ghana is generally oriented around three important features that have emerged as significant areas of shared concern. The focus is on creating capacity, increasing carbon visibility, and protecting property rights. Ghana is making remarkable strides in the REDD+ arena. In Ghana, REDD+ initiatives are planned with obvious inclusive and welcoming politics in mind. The country has included an approach 'that prioritizes the need for an accountable and participatory process with the effective participation of women and local communities' (Ghana REDD+ Strategy, 2016). REDD+ safeguards in Ghana overlook the political and competing interests of actors (Johnson, 2021). If the projects had not been undertaken, emission reductions and other benefits would not have occurred or been achieved. It is a pay-for-performance scheme, with payments postponed until the project can demonstrate the benefits of carbon stock enhancement (CSE), sustainable forest management (SFM), and conservation. Creating a REDD+ initiative or program takes time and requires a high degree of technical competence, as well as effective community engagement and large financial resources. REDD+ efforts, on the other hand, should be structured in a fair, egalitarian, and transparent manner, according to stakeholders. In contrast, if those who are making these changes are not adequately compensated based on their judgments of what is fair, reasonable, and sensible, they will be hesitant to fully engage in the project's goals, and the initiative will eventually fail. The lack of financial resources and time also limits REDD+ from gaining 'full and effective' participation from communities (Johnson, 2021).

The Congo Basin in central Africa is home to the world's second largest tropical rainforest, which offers an important source of income for the region's population. This forest is critical for Africa and the world's adaptability to climate change. Men and women have different associations with the forest, therefore their reactions to climate change and REDD+ activities in Congo may differ. In three countries, including Congo, Brown (2011) revealed that women have been underrepresented in conversations about climate change and REDD+. It will be catastrophic to attain the REDD+ goal without putting the interests and concerns of its residents, particularly women, at the forefront of the debate. Climate change, forest access, forest management, and carbon benefit sharing all have gender dimensions that must be addressed. Measures to promote the successful participation of all stakeholders are vital and must be followed. If gender consideration is not effectively integrated into future REDD+ plans and actions in Congo, future national REDD+ programs would probably fail.

The main actors of the REDD+ process in Mozambique include members of the government, academia, civil society, and the private sector. Their REDD+ method allows them to integrate climate change mitigation, biodiversity conservation, and poverty reduction. The debate centers on how indigenous peoples and local communities' rights will be respected throughout the implementation of REDD+. Ludwig (2012), depicts the challenges involved in protecting the rights of local communities in Mozambique. The study revealed that constraints regarding property and participation rights arise in particular, from incomplete procedures of delimitation and titling, corruption, lack of rights awareness, lack of democratic structures within the community, or poor infrastructure. REDD+ potential is not able to guarantee sufficient rights protection because the concentration is on the reduction of carbon emissions. He suggests the REDD+ process be equipped with substantial safeguards for indigenous people's and local communities' rights. Sitoe et al. (2012) in his study, also made a detailed examination of the key factors that can aid in the promotion of efficient, effective, and equitable REDD+ policies in Mozambique. One such equitable factor he did not mention is achieving gender justice by integrating a framework that recognizes gender issues. The study also revealed that, among other factors, agriculture and wood fuels constitute the two major causes of deforestation and forest degradation in Mozambique. These two factors are linked to the sustenance of the indigenous people's everyday livelihoods. REDD+ policies should be based on inclusive decision-making to address the question of fair representation of both sexes to ensure gender equity to achieve gender justice and safeguards.

Tanzania is one of nine pilot countries for the United Nations REDD+ Program. It receives significant funding from the Norwegian, Finnish, and German governments, and is a participant in the World Bank's Forest Carbon Partnership Facility. The interventions aim to mitigate greenhouse gas emissions, provide income to rural communities, and conserve biodiversity. The Tanzania REDD+ program is challenged with the following including currently inadequate baseline forestry data sets (needed to calculate reference emission levels), inadequate government capacity and insufficient experience in implementing REDD+-type measures at operational levels. Jacob and Brockington (2020) discovered that a key challenge facing interventions in the forestry sector in Tanzania is how to ensure that benefit sharing arrangements meet the needs and aspirations of poor rural people. The study findings showed that the majority of respondents felt benefits are concentrated on an elite group due to weak governance mechanisms. These challenges are being addressed by combined donor support to implement a national forest inventory; remote sensing of forest cover; enhanced capacity for measuring, reporting, and verification; and pilot projects to test REDD+ implementation linked to the existing Participatory Forest Management Programme. Additionally, for REDD+ to succeed, current users of forest resources must adopt new practices, including the equitable sharing of benefits that accrue from REDD+ implementation. As interest and effort build around REDD+, it is important to remember that, any other intervention, is likely to be shaped by the history of Tanzania's forestry sector, especially community-based forest management. The conclusion is that even in a country with considerable donor support, progressive forest policies, laws and regulations, an extensive network of managed forests, and an increasingly developed locally based forest management approaches, implementing REDD+ presents many challenges to achieving gender justice for the indigenous people. REDD+ is a payment-by-performance approach, with payments deferred until the project can demonstrate carbon stock enhancement.

Benefit sharing among the people will have to meet the needs and inspirations of the people and inclusive decisions have to be encouraged. This is important because it concerns the right of people. Compensation should be detailed enough to pay off people's attachment to the forest. Inclusion is important because both men and women have the same attachment to the forest. Therefore, they should be consulted and present when the compensation package is discussed. This will ensure gender equity and justice. Government mechanisms to transform political talks into clear gender action plans to facilitate inclusion and participation in REDD+ actions with strong law enforcement are the way to go for Africa. This will achieve gender equity and justice and enhance REDD+.

Although difficult to calculate systematically the benefits derived from forests, it plays a significant economic role at the continental, regional, national, and local levels in Africa.

Benefit sharing cannot be discussed and achieve fairness and equity if the access and withdrawal rights of the parties involved are not considered. These rights should be the center of the debate when sharing the benefits from the resources. Such a debate is likely to intensify tensions as actors concerned attempt to justify why they should have more access and use rights potentially evoking a defensive chain of reactions from other resource users who stand to lose their allocated rights due to re-arrangements (Soliev et al., 2017). As a result of re-arrangement activities of their forest resources, the preferences and choices of people in a particular area are changed. The rearrangements, therefore, have to cater to the new changes in their lives as well as the old privileges taken from them. In Agrawal et al. (2011), one major focus area of REDD+ is protecting indigenous people's rights. This makes sure that indigenous people are listened to, their concerns are at the table of discussion, and the appropriate benefits are allocated to them. This ensures that the indigenous people are not forced into benefit agreementswhich they are not willing to submit to. Again, this ensures that vulnerable indigenous people, especially women, are exceptionally taken care of. Gender dimensions must be addressed regarding climate change, forest access, forest management, carbon benefit sharing, and strategies to enable effective engagement of all stakeholders. This is critical in addressing more sustainable forest management and fulfilling the goals of REDD+, which aims to reduce emissions and save forests in degrading-forest countries and this requires gender parity. REDD+ is recognized in the Paris Agreement as a key mitigation policy for addressing climate change. However, a major challenge that has impeded REDD+ is ensuring gender equity (Pham et al., 2021). The Cancun Agreement of the UNFCCC required participating countries to respect gender in REDD+ design and implementation (UNFCCC, 2010). However, only a few scholars have until now addressed the relevance of gender relations within debates about the sustainable livelihoods of indigenous people and their self-determined ways of living (Löw, 2020).

Mainstreaming gender in REDD+ policies depends on a broad governance framework (encompassing institutional setting, capacity, and resources) and power structure being in place; this differs across countries and regions (Gurung et al., 2011; Ihalainen et al., 2017). REDD+ policies must aim at being gender-responsive. A gender-responsive policy takes into account and resolves gender disparities, as well as past gender prejudices. A gender-responsive policy aims to 'do better' (Blomstrom & Burns, 2015) by addressing the socio-cultural basis of gender differences and power relations in the local setting (Tyagi & Das, 2018). Addressing gender disparities by using the past as the bases to implement inclusive, better, and good policies are ensuring gender justice. That is integrating concerns of the vulnerable, especially women, into the goals, aims, and objectives of REDD+. This is so because a solution that takes into consideration the concerns of all citizens for the betterment of their livelihood is what REDD+ works toward achieving with fringe communities. It also ensures that women have equal access to financial resources and benefits, as well as equal involvement and influence in decision-making processes. It ensures again that, women have the right to point out the type and form of compensation they are willing to accept from stakeholders.

Gender justice requires an integrated gender perspective to address socio-economic, political, and ecological inequality and power imbalances as a truly socially and gender equitable policy. According to research, indigenous and other marginalized rural women have a key role in tackling climate change, reducing deforestation, and guaranteeing sustainable forest management. This is a result of specific predicaments indigenous women face. These women have an important role in forest management, for example, by engaging in traditional agroforestry and gathering firewood and non-wood forest products (NWFPs) for food, medicine, and fodder. Women work in nurseries and patrol and monitor forests in several countries, such as Indonesia, Ghana, and Vietnam. Therefore, women should be among the beneficiaries of forest-related sustainable development programs, given their engagement in forest management, yet they are frequently kept out of international and national policies when it comes to sharing the advantages of climate investment incentives and communities' decision-making. Men and women have diverse relationships with and use of the forest and hence may have varied reactions to climate change and REDD+ initiatives. Women have had little engagement in conversations about climate change or REDD+, according to research. A person's vulnerability to climate change depends in part on gender roles and relations; rural women in developing countries are one of the most vulnerable groups (IPCC, 2007). The statement is true as women in the rural settings of Africa are often dependent on natural resources for their livelihoods, engage in agricultural work, and are in charge of collecting water and fuel for their families' use.

Although gender forms an important policy component of REDD+, the funding allocated to gender is often limited and seen as an ad hoc budget line in other programs. This creates major challenges in treating gender as a concrete priority (Pham et al., 2021). If a country gives full legal recognition at both international and national levels—to gender equity, this will provide a key enabling condition for gender to be better mainstreamed in REDD+ policies. Countries are to be committed to mainstreaming gender in REDD+ by respecting international agreements like the Convention on the Elimination of All Forms of Discrimination Against Women (CEDAW), The International Labor Organization (ILO) Convention No. 169, the United Nations Declaration on the Rights of Indigenous Peoples (UNDRIP), and UNFCCC's Cancun Safeguards by all REDD+ countries (Pham et al., 2021). African countries should also have a recognized legal framework that will be responsible for planning and implementation to create a condition that makes gender mainstreaming possible, for a better understanding of women's role in forest management. There should be a further breakdown of these legal frameworks into agencies that work directly with women forest dependents and dwellers to facilitate equal participation and empowerment. This place women in a good position to make informed demands that satisfy their needs. Again, individual African countries should be able to translate these international laws into country-specific legal frameworks for them to sit well with the citizens of the country.

For African countries to effectively mainstream gender into REDD+ and to ensure gender justice, they will need technical support and financial help to make sure women's voices and concerns are heard at every level of the consultation. Research shows that seven countries in Africa (41.2%) had received support from international organizations in terms of financial and technical assistance for gender mainstreaming in REDD+ processes. This assistance came in the form of pilot REDD+ projects focused on gender issues, like the CARE/HIMA project (2010–2013) in Tanzania (Campese, 2011), gender development in the REDD+ Roadmap in Cameroon (IUCN, 2015; Gilardi & Mazet, 2019); formulation of a gender engagement/action plan in Burkina Faso (Government of Burkina Faso, 2017). Women organization groups within African countries, both governmental and non-governmental, should actively get involved in getting African women to participate in the negotiation stage to encourage mainstreaming of gender into REDD+ to ensure gender justice. An example is Mozambique's Readiness Preparation Proposal (R-PP) which placed the Ministry of Women and Social Affairs and the Women's Forum in charge of social impacts and gender equity (Government of Mozambique, 2013). Although most countries have an overall legal framework on gender, political commitment is not translated into action. The lack of policies not only means no incentives for law enforcement and the monitoring of progress and activities related to gender mainstreaming; it also fails to help local authorities, who often lack the skills and gender awareness to translate these political visions into practical action (Pharm et al., 2016).

4 Conclusion

In conclusion, this chapter discovered in the literature that benefit sharing promotes inclusiveness, fairness, and transparency among forest-dependent communities of Africa in the context of climate change and REDD+. This is seen as a strategy for mitigating climate change and achieving the objective of REDD+. It however underscores difficulties in calculating systematically, the right size of the forest, socio-cultural systems in a variety of ways, uninformed consent, food insecurity, local people's livelihoods, women's policies, and the satisfaction the people derive from the forest to be monetized. This creates a problem for the international players, thereby affecting performance and stakeholders' involvement in the implementation of successful forest-protected projects by incorporating forest-dependent women of Africa. African countries need to integrate gender into all environmental policies including REDD+. African women are also to be supported and empowered to effectively participate in their concerns to be addressed. It stood out that, the discussion on forest loss and damage caused by climate change and how to address and compensate it for the associated damage to livelihoods and properties provide conservation of the forest as a great opportunity for the world to fight climate change. Governments of African countries should look back as well as forward to highlight lessons learned and best practices from traditional indigenous knowledge for integration into the various REDD+ policies for inclusiveness. This will be obligatory on the government as well as the communities in which women will be included. We do also recommend the strengthening and integration of gender issues in future research investments. That is more concentration of women's concerns at the center of drafting future forest investment documents. The presentation of these recommendations for designing and implementing projects that lead to more gender equitable outcomes will also contribute to the realization of larger program objectives to strengthen women's and men's resilience.

We agree with Bayrak and Marafa's (2016) suggesting that REDD+ implementation should take a holistic approach that takes into account the often complex relationships between forest-dependent communities and the natural environment, including the incorporation of traditional forest management systems and the provision of viable alternatives to the loss of agricultural land. This holistic approach of incorporation of forest-dependent communities should pay more attention to women whose major livelihood activities depend on the natural environment. In the provision of viable alternatives to loss of agricultural lands, although women in Africa may not have hectares of lands registered in their names, the policy should still look at granting the women better livelihood alternative sources.

REDD+ is currently the most important finance-driven climate mechanism in forest governance (Soliey et al., 2021). Indigenous and other oppressed rural women play a critical role in addressing climate change, decreasing deforestation, and ensuring sustainable forest management, according to intersectional feminist ideas. To achieve fully the focus and goal of REDD+ in African countries, each country needs to have its national REDD+ policy. This policy will be a check to the regional policy to make sure that individual countries are working in the same direction. Individual countries should pay particular attention to the REDD+ rules, components, activities, programs, and projects. This is to ensure that, each particular country's policy is aligned with the African regional policy.

The sustainable management of the forest provides essential carbon sinks; for example, the Congo Basin absorbs 1.2 billion tons of CO_2 each year (De Wasseige et al., 2015). REDD+ will be able to meet its goal if more funding and resources are committed to paying the increased costs at the speed and level required. An inclusive approach to forest management, equitable benefit sharing and mainstreaming gender into forest management are the ways REDD+ should go. This is an approach to forest management where all the concerns of the various parties who have a direct dependence on the forest are incorporated. REDD+ implementers should develop communication channels with the local inhabitants, especially the vulnerable whom the majority are women, on what economic livelihood will keep them away from the forest than consulting ministries and environmentalists who have not even stepped foot in these fringe communities. The locals who already have indigenous knowledge should be further trained in climate-smart agricultural technologies which reduce land use pressures by increasing the productivity of food crops and other raw materials. This has the power to reduce the wastage of agricultural lands. This is so because a small piece of land with the necessary climate-smart technology can provide for the family's food and financial needs throughout the year. This approach has the potential to prevent locals from encroaching on forest reserve lands. REDD+ programs should have the policy to compensate the fringe communities with food items and other co-benefits to balance the losses caused by climate change.

Appendix

See Table 4.

Table 4 Key literature description on climate change, REDD+, and gender justice

Description of climate change, REDD+, and gender justice	*References*
1. Urban planning and sustainable development: The inadequacy and paucity of reliable records on climate change adaptation to accurately measure the impact and effectiveness of planning interventions in meeting and contributing to the SDGs	Addaney and Cobbinah (2019)
2. Forests, poverty, and equity: This paper places forestry in the larger context of rural development and therefore in the current debates on poverty and inequality for natural assets to contribute to environmental, economic, and empowerment outcomes	Anderson et al. (2006)
3. Devolution of rights: In a long series of conservation fads, recognizing fads and thinking of them as learning opportunities is part of accepting that the practice of conservation has a culture, just like all other practices	Angelsen et al. (2017)
4. Impact of REDD+: provides an in-depth analysis of the impact of REDD+ on forest-dependent communities. The dimensions which are included in this review are institutions and governance, livelihoods, socio-cultural aspects, and the environment	Bayrak and Marafa (2016)
5. Biodiversity in the intact landscape: shows that some risks were disproportionately high in relatively intact landscapes; even minimal deforestation has had severe consequences for vertebrate biodiversity. Founding showed little support for the alternative hypothesis that forest loss is most detrimental in already fragmented landscapes	Betts et al. (2017)
6. Resilience for adaptation: The notion of resilience was questioned from very different angles, confronting concepts, specific risk management strategies, case studies and national policies, from different perspectives, biophysical, economic, or social and institutional, and at various scales, from farm and household to national and global	Braatz (2012)

(continued)

Table 4 (continued)

Description of climate change, REDD+, and gender justice	*References*
7. Responses to global warming: Examining multi-dimensional postures may aid in understanding historical and ongoing responses to global warming. Reflecting attention on fossil fuel products	Bonneuil et al. (2021)
8. Responses of carbon-degrading: Warming effects on ligninase activity increased with experiment duration. The results suggest that soil microorganisms sustain long-term increases in SR with warming by gradually increasing the degradation of the recalcitrant carbon pool. Increases in ligninase activity were positively correlated with changes in SR, while no such relationship was found for cellulase	Chen et al. (2018)
9. Women's negotiation power: Women are not normally family heads and as such not custodians of the land as well as the forest resources in Ghana	Derkyi (2012)
10. Content analysis: Analyzing textual data to identify and aggregate framings, patterns, constructs, and expressions to address a research problem or question	Krippendorff (2018)
11. Gender equality: Valuable method for assessing gender equality in a range of diverse but related domains across numerous disciplines	Schmidt and Graversen (2020)
12. Data collection: observed that systematic mapping collates, describes, and catalogues available data directly related to a question of interest	James et al. (2016)
13. Mitigation approach: climate change mitigation approach was conceptualized to incentive developing countries to reduce carbon emissions from deforestation and forest degradation by providing compensatory payments for measurable and verifiable CO_2 emissions	Sheng et al. (2018)
14. Importance of the environment: Is essential for their survival, and forms an important and integral part of many forest-dependent and indigenous communities' livelihoods, institutions, cultures, social relations, and identities	Isyaku (2021)

(continued)

Table 4 (continued)

Description of climate change, REDD+, and gender justice	*References*
15. Benefit-sharing: established that in the country, four different types of benefit sharing arrangements within the forest sector have been identified to achieve viable REDD+ projects. These include; Constitutional Timber Revenue benefit sharing, Modified Taungya System benefit sharing, Commercial Plantation benefit-sharing, and Community Resource Management Area (CREMA) benefit sharing	Dumenu et al. (2014)
16. Women participation in REDD+ in Congo: revealed that women have been underrepresented in conversations about climate change and REDD+	Brown (2011)
17. Benefit sharing in Mozambique: A detailed examination of the key factors that can aid in the promotion of efficient, effective, and equitable REDD+ policies in Mozambique. One of such equitable factors he did not mention is achieving gender justice by integrating framework that recognizes gender issues	Sitoe et al. (2012)
18. Rights of local communities: Depicts the challenges involved in protecting the rights of local communities in Mozambique. The study revealed that constraints regarding property and participation rights arise, in particular, from incomplete procedures of delimitation and titling, corruption, lack of rights awareness, lack of democratic structures within the community or poor infrastructure	Ludwig (2012)
19. Relevance of gender: only a few scholars have addressed the relevance of gender relations within debates about the sustainable livelihoods of indigenous people and their self-determined ways of living	Löw (2020)
20. Forest dependency: FAO Report (2000) in the year 2000, 300 million people directly depended on the forest	FAO Report (2000)
21. Rewards in REDD+: There is a shift toward results-based rewards in REDD+. For example, Phase 3 transfers funds only after countries implement Phases 1 and 2 and reductions in greenhouse gases (GHG) are verified	Lujan-Moreno et al. (2018)

References

Abebe, M. A. (2014). Climate change, gender inequality and migration in East Africa. *Washington Journal of Environmental Law & Policy, 4*(1), 104–140.

Addaney, M. (2018). The law as shelter: The interface between women and climate change adaptation responses in Africa. *Journal of Comparative Law in Africa, 5*(2), 117–148.

Addaney, M., & Cobbinah, P. B. (2019). Climate change, urban planning and sustainable development in Africa: The difference worth appreciating. In P. B. Cobbinah & M. Addaney (Eds.), Geography of climate change adaptation in Urban Africa. Springer. https://doi.org/10.1007/978-3-030-04873-0

Addaney, M., & Moyo, C. G. (2018). Women's rights, gender and climate change law in Africa: Advancing an equity agenda. *Journal of Law, Society and Development, 5*(1), 18-pages.

Agrawal, D., Bernstein, P., Bertino, E., Davidson, S., Dayal, U., Franklin, M., ... & Widom, J. (2011). *Challenges and opportunities with big data 2011-1.*

Angelsen, A., Brockhaus, M., Duchelle, A. E., Larson, A., Martius, C., Sunderlin, W. D., &Wunder, S. (2017). Learning from REDD+: A response to Fletcher et al. *Conservation Biology, 31*(3), 718–720.

Autio, A., Johansson, T., Motaroki, L., Minoia, P., & Pellikka, P. (2021). Constraints for adopting climate-smart agricultural practices among smallholder farmers in Southeast Kenya. *Agricultural Systems, 194*, 103284. https://doi.org/10.1016/j.agsy.2021.103284

Babugura, A. (2010). *Gender and climate change: South Africa case study.* Heinrich Böll Foundation Southern Africa.

Balama, C., Augustino, S., Eriksen, S., & Makonda, F. B. S. (2016). Forest adjacent households' voices on their perceptions and adaptation strategies to climate change in Kilombero District, Tanzania. *SpringerPlus*, 5https://doi.org/10.1186/s40064-016-2484-y

Bayrak, M. M., & Marafa, L. M. (2016). Ten years of REDD+: A critical review of the impact of REDD+ on forest-dependent communities. *Sustainability, 8*(7), 620.

Becerra, S., Saqalli, M., Gangneron, F., & Dia, A. H. (2016). Everyday vulnerabilities and "social dispositions" in the Malian Sahel, an indication for evaluating future adaptability to water crises? *Regional Environmental Change, 16*(5), 1253–1265.

Betts, M. G., Wolf, C., Ripple, W. J., Phalan, B., Millers, K. A., Duarte, A., Butchart, S. H. M., & Levi, T. (2017). Global forest loss disproportionately erodes biodiversity in intact landscapes. *Nature, 547*(7664), 441–444.

Blomstrom, E., & Burns, B. (2015). Global policy landscape: A supporting framework for gender-responsive action on climate change. *Roots for the future: The landscape and way forward on gender and climate change* (pp. 47–80).

Bonneuil, C., Choquet, P. L., & Franta, B. (2021). Early warnings and emerging accountability: Total's responses to global warming, 1971–2021. *Global Environmental Change, 71*, 102386.

Bowen, G. A. (2009). Document analysis as a qualitative research method. *Qualitative Research Journal*.

Braatz, S. (2012). Building resilience for adaptation to climate change through sustainable forest management. *Building Resilience for Adaptation to Climate Change in the Agriculture Sector, 23*, 117.

Brockhaus, M., Wong, G., Luttrell, C., Loft, L., Pham, T. T., Duchelle, A. E., & Di Gregorio, M. (2014). *Operationalizing Safeguards in National REDD+ Benefit-sharing Systems: Lessons on effectiveness, efficiency and equity* (Vol. 2). CIFOR.

Brown, H. C. P. (2011). Gender, climate change and REDD+ in the Congo Basin forests of Central Africa. *International Forestry Review, 13*(2), 163–176.

Caroli, G., Tavenner, K., Huyer, S., Sarzana, C., Belli, A., Elias, M., Pacillo, G., & Läderach, P. (2022). *The gender-climate-security nexus: Conceptual framework, CGIAR portfolio review, and recommendations towards an agenda for one CGIAR* (Position Paper No. 2022/1). CGIAR FOCUS Climate Security.

Counsell, S. (2009). *Forest governance in Africa*.

Chen, J., Luo, Y., García-Palacios, P., Cao, J., Dacal, M., Zhou, X., Li, J., Xia, J., Niu, S., Yang, H., Shelton, S., Guo, W., & van Groenigen, K. J. (2018). Differential responses of carbon-degrading enzyme activities to warming: Implications for soil respiration. *Global Change Biology, 24*(10), 4816–4826.

Cook, J., Oreskes, N., Doran, P. T., Anderegg, W. R., Verheggen, B., Maibach, E. W., Carlton, J. S., Lewandowsky, S., Skuce, A. G., Green, S. A., Nuccitelli, D., Jacobs, P., Richardson, M., Winkler, B., Painting, R., & Rice, K. (2016). Consensus on consensus: A synthesis of consensus estimates on human-caused global warming. *Environmental Research Letters, 11*(4), 048002.

Dalsgaard, S., Funder, M., Hagelberg, N., Harrison, P., Haule, C., Kabalimu, K., & Danielsen, F. (2010). *Getting ready for REDD+ in Tanzania: A case study of progress and challenges*.

Deressa, T. T., Hassan, R. M., Ringler, C., Alemu, T., & Yesuf, M. (2009). Determinants of farmers' choice of adaptation methods to climate change in the Nile Basin of Ethiopia. *Global Environmental Change, 19*, 248–255.

Derkyi, M. A. A. (2012). *Fighting over forest: Interactive governance of conflicts over forest and tree resources in Ghana's high forest zone* (No. 41). African Studies Centre.

De Wasseige, C., Tadoum, M., Atyi, E. A., & Doumenge, C. (2015). *The forests of the Congo Basin-Forests and climate change*.

Dillon, A., Mueller, V., & Salau, S. (2011). Migratory responses to agricultural risk in northern Nigeria. *American Journal of Agricultural Economics, 93*(4), 1048–1061.

Diouf, N. S., Ouedraogo, I., Zougmoré, R. B., Ouedraogo, M., Partey, S. T., & Gumucio, T. (2019). Factors influencing gendered access to climate information services for farming in Senegal. *Gender, Technology and Development, 23*(2), 93–110.

Doss, C., Deere, C. D., Oduro, A. D., & Suchitra, J. Y. (2012). *The rural gender asset and wealth gaps: Evidence from Ghana, Ecuador, Uganda and Karnataka, India*. Indian Institute of Management, Bangalore.

Dumenu, W. K., Samar, S., Mensah, J. K., Derkyi, M., Oduro, K. A., Pentsil, S., ... & Obeng, E. A. (2014). Benefit sharing mechanism for REDD+ implementation in Ghana. Consultancy Report. *Forestry Commission, Accra, Ghana*.

FAO. (2000). *Global forest resources assessment 2000*. Rome

FAO. (2002). *2nd expert meeting on harmonizing forest-related definitions for use by various stakeholders*. Rome

FAO. (2011). *The state of food and agriculture: Women in agriculture closing the gender gap for development*. FAO.

FAO. (2015). *Climate change and food security: Risks and responses*. FAO.

Fisher, M., & Carr, E. R. (2015). The influence of gendered roles and responsibilities on the adoption of technologies that mitigate drought risk: The case of drought-tolerant maize seed in eastern Uganda. *Global Environmental Change, 35*, 82–92.

Fuhrman, S., Janoch, E., Koch, R., Oo, K., Parra, V., & Rawe, T. (2020). *Left out and left behind: Ignoring women will prevent us from solving the hunger crisis*. CARE Policy Report.

Gebrehiwot, T., & van der Veen, A. (2013). Farm level adaptation to climate change: The case of farmer's in the Ethiopian Highlands. *Environmental Management, 52*, 29–44.

Ghana REDD+ Strategy, Republic of Ghana. (2016). https://www.oldwebsite.fcghana.org/userfiles/files/REDD%2B/Ghana%20REDD%2B%20Strategy.pdf

Gilardi, K. V., & Mazet, J. A. (2019). The United States agency for international development emerging pandemic threats PREDICT project—Global detection of emerging wildlife viral zoonoses. *Fowler's Zoo and Wild Animal Medicine Current Therapy, 9*, 110.

Gizachew, B., Astrup, R., Vedeld, P., Zahabu, E. M., & Duguma, L. A. (2017). REDD+ in Africa: Contexts and challenges. *Natural Resources Forum*. https://doi.org/10.1111/1477-8947.12119

González, J. B., & Sánchez, A. (2022). Multilevel predictors of climate change beliefs in Africa. *PLoS ONE, 17*(4), e0266387.

Gray, C. L., & Mueller, V. (2012). Drought and population mobility in rural Ethiopia. *World Development, 40*(1), 134–145.

Hallema, D. W., Sun, G., Caldwell, P. V., Norman, S. P., Cohen, E. C., Liu, Y., Bladon, K. D., & McNulty, S. G. (2018). Burned forests impact water supplies. *Nature Communications, 9*(1), 1–8.

Henriksson, R., Vincent, K., Archer, E., & Jewitt, G. (2020). Understanding gender differences in availability, accessibility and use of climate information among smallholder farmers in Malawi. *Climate and Development*. https://doi.org/10.1080/17565529.2020.1806777

International Forestry Review Vol.13 (2), 2011 163 Gender, climate change and REDD+ in the Congo Basin forests of Central Africa H.C. PEACH BROWN Environmental Studies, University of Prince Edward Island, Charlottetown, PEI, Canada, C1A 4P3.

International Union for Conservation of Nature Report, 2015.

IPCC. (2007). *Report of the nineteenth session of The Intergovernmental Panel on Climate Change (IPCC) Geneva*, 17–20 (a.m. only) April 2002.

Isyaku, U. (2021). What motivates communities to participate in forest conservation? A study of REDD+ pilot sites in Cross River, Nigeria. *Forest Policy and Economics, 133*, 102598.

Jacob, T., & Brockington, D. (2020). Learning from the other: Benefit sharing lessons for REDD+ implementation based on CBFM experience in Northern Tanzania. *Land Use Policy, 97*, 103010.

James, S., Herman, J., Rankin, S., Keisling, M., Mottet, L., & Anafi, M. A. (2016). *The report of the 2015 US transgender survey.*

Johnson, S. (2021). Discourse and practice of REDD+ in Ghana and the expansion of state power. *Sustainability, 13*(20), 11358.

Kakota, T., Nyariki, D., Mkwambisi, D., & Kogi-Makau, W. (2011). Gender vulnerability to climate variability and household food insecurity. *Climate and Development, 3*(4), 298–309.

Kalaba, F. K., Quinn, C. H., & Dougill, A. J. (2013). The role of forest provisioning ecosystem services in coping with household stresses and shocks in Miombo woodlands, Zambia. *Ecosystem Services, 5*, 143–148.

Krippendorff, K. (2018). *Content analysis: An introduction to its methodology.* Sage.

Kuyper, J., Schroeder, H., & Linnér, B. O. (2018). The evolution of the UNFCCC. *Annual Review of Environment and Resources, 43*, 343–368.

Lama, P., Hamza, M., & Wester, M. (2021). Gendered dimensions of migration in relation to climate change. *Climate and Development, 13*(4), 326–336.

Larson, A. M., Dokken, T., & Duchelle, A. E. (2016). Can safeguards guarantee gender equity? Lessons from research on women in early REDD+ implementation. *REDD+ Safeguards Brief, 4*, CIFOR.

Larson, A. M., Solis, D., Duchelle, A. E., Atmadja, S., Resosudarmo, I. A. P., Dokken, T., & Komalasari, M. (2018). Gender lessons for climate initiatives: A comparative study of REDD+ impacts on subjective wellbeing. *World Development, 108*, 86–102.

Löw, C. (2020). Gender and indigenous concepts of climate protection: A critical revision of REDD+ projects. *Current Opinion in Environmental Sustainability, 43*, 91–98.

Ludwig, G. (2012). *Foundations of quantum mechanics I*. Springer Science & Business Media.

Lujan-Moreno, G. A., Howard, P. R., Rojas, O. G., & Montgomery, D. C. (2018). Design of experiments and response surface methodology to tune machine learning hyperparameters, with a random forest case-study. *Expert Systems with Applications, 109*, 195–205.

Mekonnen, Z. (2022). Intra-household gender disparity: Effects on climate change adaptation in Arsi Negele district, Ethiopia. *Heliyon, 8*, e08908. https://doi.org/10.1016/j.heliyon.2022.e08908

McOmber, C. (2020). Women and climate change in the Sahel (West African Papers, No. 27). OECD Publishing https://doi.org/10.1787/e31c77ad-en.

Nagel, J. (2015). *Gender and climate change: Impacts, science, policy*. Routledge.

Nkhata, B. A., Mosimane, A., Downsborough, L., Breen, C., & Roux, D. J. (2012). A typology of benefit sharing arrangements for the governance of social-ecological systems in developing countries. *Ecology and Society, 17*(1).

Nurse, M. S., & Grant, W. J. (2020). I'll see it when I believe it: Motivated numeracy in perceptions of climate change risk. *Environmental Communication, 14*(2), 184–201.

Pham, T. T., Tran, N. L. D., Nong Nguyen Khanh, N., & Nguyen Dinh, T. (2021). Mainstreaming gender in REDD+ policies and projects in 17 countries. *Journal of Environmental Policy & Planning*, 1–15.

Pharm, S. K. B., Pharm, Y. N. B., Pharm, S. B. B., Rattanaumpawan, P., Boonyasiri, A., & Thamlikitkul, V. (2016). Epidemiology of antibiotic use and antimicrobial resistance in selected communities in Thailand. *Journal of the Medical Association of Thailand, 99*(3), 270–275.

Revelo, L. A. (2021). *Gender equality in the midst of climate change: What can the region's machineries for the advancement of women do?* Gender Affairs series, No. 159 (LC/TS.2021/79), Santiago, Economic Commission for Latin America and the Caribbean (ECLAC).

Schmidt, E. K., & Graversen, E. K. (2020). Developing a conceptual evaluation framework for gender equality interventions in research and innovation. *Evaluation and Program Planning, 79*, 101750.

Sellers, S. (2016). *Gender and climate change: A closer look at existing evidence*. Global Gender and Climate Alliance.

Setyowati, A. (2012). Ensuring that women benefit from REDD+. *Unasylva, 239*(63), 57–62.
Sitoe, A., Salomão, A., & Wertz-Kanounnikoff, S. (2012). *The context of REDD+ in Mozambique: Drivers, agents and institutions* (Vol. 79). CIFOR.
Skinner, E. (2011). *Gender and climate change overview report*. BRIDGE Cutting Edge Pacon Gender and Climate Change, Institute of Development Studies, Brighton.
Soliev, I., Theesfeld, I., Abert, E., & Schramm, W. (2021). Benefit sharing and conflict transformation: Insights for and from REDD+ forest governance in sub-Saharan Africa. *Forest Policy and Economics, 133*, 102623.
Soliev, I., Theesfeld, I., Wegerich, K., & Platonov, A. (2017). Dealing with "baggage" in riparian relationship on water allocation: A longitudinal comparative study from the Ferghana Valley. *Ecological Economics, 142*, 148–162.
Soliev, I., & Theesfeld, I. (2017). Reframing for sustainability: Exploring transformative power of benefit sharing. *Sustainability, 9*(8), 1486.
Stevenson, E. G. J., Ambelu, A., Caruso, B. A., Tesfaye, Y., & Freeman, M. C. (2016). Community water improvement, household water insecurity, and women's psychological distress: An intervention and control study in Ethiopia. *PLoS ONE, 11*(4), e0153432. https://doi.org/10.1371/journal.pone.0153432
Susskind, L. E., & Ali, S. H. (2014). *Environmental diplomacy: Negotiating more effective global agreements*. Oxford University Press.
Tadesse, D. (2010). *The impact of climate change in Africa*. Institute for Security Studies.
The United States Agency for International Development Report, 2014.
Tyagi, N., & Das, S. (2018). Assessing gender responsiveness of forest policies in India. *Forest Policy and Economics, 92*, 160–168.
Twyman, J., Green, M., Bernier, Q., Kristjanson, P., Russo, S., Tall, A., Ampaire, E., Nyasimi, M., Mango, J., McKune, S., Mwongera, C., & Ndourba, Y. (2014). *Adaptation actions in Africa: Evidence that gender matters* (CCAFS Working Paper No. 83). Copenhagen. CGIAR Research Program on Climate Change, Agriculture and Food Security (CCAFS).
United Nations Framework Conversing on Climate Change Report, 2010.
United Nations Development Programme, UNDP. (2011). *Global gender and climate alliance, overview of linkages between gender and climate change*. http://www.undp.org/content/dam/undp/library/gender/Gender%20and%20Environment/PB1-APOverview-Gender-and-climate-change.pdf
UNDP. (2016). *Gender and climate change: Gender and REDD+*. Policy Brief 6.

UN WOMEN. (2021). *Women's representation in local government: A global analysis* (Working Paper). UN WOMEN, governance and participation section.

Woittiez, L. S., Rufino, M. C, Giller, K. E., & Mapfumo, P. (2013). The use of woodland products to cope with climate variability in communal areas in Zimbabwe. *Ecology and Society, 18*(4). https://doi.org/10.5751/ES-05705-180424

World Bank. (2019). *Gender mainstreaming in REDD+ processes and programs in Africa (2019–2025)—An initial strategy*. World Bank.

World Bank. (2021). *Gender equity in land and forest tenure in REDD+ programming: Deep dive country profiles.*

Wornell, E. J., Tickamyer, A. R., & Kusujiarti, S. (2015). Gender mainstreaming principles in Indonesia's REDD+ program: A document analysis. *Journal of Sustainable Development, 8*(8), 159.

Zakari, S., Ying, L., & Song, B. (2014). Factors influencing household food security in West Africa: The case of Southern Niger. *Sustainability, 6*, 1191–1202.

CHAPTER 6

Multi-Level Climate Change Adaptation Governance in Lagos in Nigeria and the Western Cape in South Africa

Olubunmi A. Afinowi

1 Introduction

The impacts of climate change will be felt at local levels, and therefore, renders the active participation of stakeholders at local, national and regional levels critical for the advancement of adaptation decisions reached through United Nations Framework Convention on Climate Change (UNFCCC) negotiations (Articles 2 & 3 UNFCCC, 1992; Dellmuth & M, 2021; Stern, 2007). This reality raises the question of governance within the climate change adaptation agenda. The all-pervasive impacts of climate change make it essential that climate action should entail the participation of all stakeholders, including the government at all levels (Ishtiaque, 2021). Bearing in mind that the guidelines on climate action are derived from global negotiations and agreements,

O. A. Afinowi (✉)
Department of Private and Property Law, University of Lagos, Lagos, Nigeria
e-mail: oafinowi@unilag.edu.ng

M. Addaney et al. (eds.), *Climate Change in Africa*,
https://doi.org/10.1007/978-3-031-30050-9_6

it is necessary to examine the extent to which these agreements are implemented at the national and sub-national levels, as well as the vertical and horizontal collaboration in governance to actualise adaptation to climate change. It is acknowledged that both state and non-state actors play major roles in carrying out adaptation to climate change (Huitema et al., 2016), however, the role of the government is vital in harnessing and laying out guidelines and legal frameworks for such cooperative climate action (Bauer, 2014). Further, proper coordination and cooperation must exist between levels of government as state actors and non-state actors to achieve a sound framework for climate adaptation. According to Bauer (2014), national governments should be more concerned with raising awareness and laying adaptation guidelines as well as providing funding, while governments at the lower levels would be more involved in implementing the adaptation process, considering they are closer to the grassroots where adaptation matters most of all (Bauer, 2014; Corfee-Morlot et al., 2009; Madzwamuse, 2011; Taylor, 2016).

In line with the Paris Agreement and the recently concluded COP26, this study examines the means and necessity of collaborative action between international, regional, national and sub-national players to alleviate the various challenges of climate change. Also, to be considered are the likely challenges and limitations that exist in applying global agendas at the grassroot levels, such as lack of technical or financial capacity, local situations and peculiarities and so on. This aspect of the research is premised on the supposition that while climate change is viewed as a global challenge, it lies with individual governments to determine suitable adaptation tracks that address specific vulnerabilities within their countries (Bauer, 2014). This chapter sets out to consider the extent of adaptation to climate change in both Nigeria and South Africa. It entails an analysis of policies on adaptation to climate change in Nigeria and South Africa. In general, the study considers adaptation in climate action within the larger confines of governance in Cape Town, South Africa and Lagos, Nigeria. The chapter discusses the policy documents on climate change at the international level and in both countries under consideration at various levels and spheres of government. The research finds that adaptation action in South Africa is more collaborative as well as cross-sectoral and encompasses disaster risk management as viable means of building resilience. Nigeria has not fully integrated these principles into its climate action and is lacking in the management of climate risks and hazards. The chapter identifies the gaps in the climate strategies of both Western

Cape and Lagos and what can be done to promote more viable adaptation action. The aim of this research is to arrive at a workable structure for interactions in the governance of climate adaptation in Africa.

2 The Nature of Multi-Level Governance and Its Relevance to Climate Action

Governance is noted to mean "the entirety of regulations – that is, the processes by which norms, rules and programs are monitored, enforced and adapted, as well as the structures in which they work – put forward with reference to solving a specific problem or providing a common good" (Zurn, 2010). Further, governance is seen as encompassing all the players, structures and processes involved in the regulating of specific interests, as distinct from other interests or endeavours (Zurn, 2010). The existence of processes and structures implies the presence of players and a collection of necessary actions which together form the governance structure. According to Zurn, Walti and Enderlein, the actions comprise negotiations, lobbying, coalitions, persuasions and threats (penalties imposed by the command-and-control system which characterises law) (Zürn et al., 2010). These processes also presuppose the existence of an array of different players including international players, government actors and the national and sub-national levels and other actors in the public and private spheres. The array of players implies collaborations at various levels, depending on factors such as jurisdiction, autonomy, financing, public participation and so on, giving rise to a system of multi-level governance.

Multi-level governance entails collaboration and partnership between various players, decision-makers and stakeholders in relation to the process of governance of collective interest or good (Zürn et al., 2010). Primarily, environmental governance (as a percussor to climate governance) emerged to tackle difficult environmental challenges by engendering environmental strategies and policies to address complex environmental issues of climate change, energy access and conservation, management of natural resources and so on (Janicke & Jorgens, 2006). Further, environmental governance embraces a global and multi-level approach due to the transboundary and inter-jurisdictional nature of environmental issues. This approach engenders collaborative action and coherence between various heterogeneous entities such as international organisations, national and sub-national governments, civil society organisations and market forces (Bridge & Perreault, 2009; Janicke & Jorgens, 2006).

It has been noted that while mitigation action is more sophisticated, adaptation action is less streamlined and entails diverse interventions, constraints and an array of stakeholders and implementation agencies (MISTRA Urban Futures). For this reason, climate adaptation entails multi-level governance with collaboration engendered between the UNFCCC, the subsequent Agreements, Protocols and Pacts and related global Agendas such as the Sustainable Development Goals and the Framework on Disaster Risk Reduction; and how these international agendas impact or shape regional and national policies on climate change; and in turn how these frame policies and action at the sub-national levels. The system also includes the participation or partnerships with private stakeholders at various levels. Given this background on governance, the chapter examines the multi-level nature of climate governance in the already highlighted case studies.

2.1 *Levels of Governance in the Constitutions of South Africa and Nigeria*

According to Hooghe and Marks (2001) requires levels of jurisdictional competence between spheres or levels of government. The Constitutions of both countries make provisions for these jurisdictional competences, as well as interactions and flow of authority between organs and institutions of government. In both countries under consideration, policy and legislative competency is shared amongst the tiers or spheres of government in the Constitution, with each level or sphere of government having its stipulated obligations and authorities (section 40, Constitution of the Republic of South Africa Act 108 of 1996) (section 3 & 4 1999 Constitution of the Federal Republic of Nigeria.).

Under the South African constitution, the spheres of government according to the Constitution are "...distinctive, interdependent and interrelated" and they together have a duty to "secure the well-being of the people" without interfering with the jurisdiction of other spheres of governments (section 41(1) (b), (f) and (g) of the Constitution of the Republic of South Africa). While the Nigerian Constitution does not make express provisions for cooperative governance, it provides for the division of powers between the recognised levels of government and empowers states to make laws for the peace, order and good governance in relation to all residual matters not within the competency of the National Assembly (section 4 1999 Constituion of Nigeria). Such residual matters as outlined in the second schedule to the 1999 Constitution includes matters pertaining to the development of individual states (Part 2, Second Schedule, 1999 Constitution of Ngeria), while the regulation of the environment falls within the jurisdictional purview of the federal government (section 20, 1999 Constitution of Nigeria).

It is noted that adaptation to climate change, raises issues of interactions and cooperation amongst levels or spheres of government and how these interactions should be coordinated. As seen above, in Nigeria environmental matters fall within the purview of the federal government, while aspects of development fall within the purview of the local and state governments; aspects of infrastructural development fall within the purview of the federal government only regarding power supply and other infrastructure that are interstate (Second Schedule, 1999 Constitution). In South Africa, there exists a better structure for government cooperation amongst the spheres of government. Whether climate change is perceived as a developmental or environmental challenge, there remains the need for inter-governmental collaboration and cooperation, the analysis of Adaptation in the Chapter highlights the level of government cooperation in this regard.

3 Theoretical Foundations of Climate Adaptation Framework

While there are several theoretical justifications for climate change adaptation governance, this chapter draws from the works of Hatfield-Dodds et al. (2007), Chaffin et al. (2014) and Schultz et al. (2015). Hatfield-Dodds et al. (2007) theorise adaptive governance from an institutional perspective where there is a duty of care to manage and appropriate tangible and intangible shared assets as regards natural resources, the environment and ecosystem biodiversity so that the shared assets can continue to serve humanity. There is a need for a collaborative approach involving all stakeholders led by the public sector based on global frameworks to address climate-prone challenges. In the same vein, it has been argued that the institutional governance approach to climate change should be innovative to address specific cases in contrast to adapting one-size fits all cases (Patterson & Huitema, 2019). This argument is premised on "unfolding uncertainties, dynamics, and pressures of climate change" (Patterson & Huitema, 2019, p. 375). These uncertainties in relation to climate change impacts make it imperative institutions to be proactive and dynamic in their multi-level governance approaches to climate action, as a way of mitigating the adverse effect of climate change. The role of the government in adaptation governance is further justified by Schultz et al., (2015, p. 7369), who notes that "adaptive governance for management of ecosystems and their services in real social-ecological landscapes … has illuminated the intricate interplay, of individual actors, social networks, organizations, and institutions, that enables or hinders societies to nurture natural capital and implement ecosystem management".

In recent years, there has been an increase in States' development of regulatory frameworks and policies on climate action (Massey et al., 2014; Nash et al., 2021). While legal frameworks have not gained so much traction, climate policies are quite common with a number of African countries adopting policies on action plans for both mitigation and adaptation. Research shows that innovations in research could be as a result of either internal or external determinants (Jordan & Huitema, 2014b). While it may be difficult to decipher the exact determinants of policy and legislation innovations in both countries, there is evidence that climate action in Africa is driven internally by the need to build climate resilience across key sectors (UNDP Climate Promise, 2022). External drivers also play a fundamental role in climate policy or regulation formulation. For instance, the UNFCCC remains the foundational driver of most climate action across participating states (Massey et al., 2014), also states with socio-economic, political or geographical similarities tend to influence each other in policy or regulation formulation (Jordan & Huitema, Innovations in climate policy: the politics of invention, diffusion, and evaluation, 2014b; Massey et al., 2014; Nash et al., 2021).

In discussing innovations in climate adaptation frameworks, the role of the national government as a major actor must be acknowledged (Jordan & Huitema, Innovations in climate policy: conclusions and new directions, 2014a). While lower levels of government may be involved in various collaborations to foster climate adaptation, the national governments play a vital role in setting the stage for implementation, collaboration with global organisations and providing the administrative resources, legal authority, and most fundamental, financing (Jordan & Huitema, 2014a). The role of the central or national government is further iterated by the fact that it acts as a steering force "to develop and project long-term visions, stimulate and oversee local approaches, and carry forward ambitious climate programs" (Jordan & Huitema, Policy innovation in a changing climate: Sources, patterns and effects, 2014c). Consequently, the research duly considers the national framework in both case studies as legitimacy proceeds from national legislation down to the sub-national.

While acknowledging the role of national governments, it is essential, as part of multi-level governance, that local urban authorities play a key role in addressing climate adaptation (MISTRA Urban Futures) other areas. The role of other stakeholders in the innovation of climate policies is also noted. Primarily, civil societies and other private players are identified as key facilitators of climate action and regulation (Nash et al., 2021) (Jordan & Huitema, Policy innovation in a changing climate: Sources, patterns and effects, 2014c). While it may be difficult to evaluate or measure the impacts of these frameworks on adaptation in the case studies ((Nash, Torney, & Matti, 2021), it is essential to acknowledge the structure and procedure that they bring to the adaptation process as will be evident in the examination of the chosen case studies in this chapter.

4 International Background for Climate Change Adaptation in South Africa and Nigeria

South Africa and Nigeria as member states of the UNFCCC and the Paris Agreement are required to integrate climate change considerations into social, economic and environmental policies to reduce adverse impacts on environmental and socio-economic conditions of life (The Paris Agreement, 2015; United Nations Framework Convention Framework on Climate Change, 1992). The Convention also calls on states to create awareness, spread the knowledge on climate change and to encourage full participation in the process of combating climate change and its impacts (Art 4 (1)(i) UNFCCC, 1992). Specifically, regarding public awareness and participation in climate change action, State members are mandated to promote inter-governance cooperation across all levels and make information on climate change readily accessible to members of the public and encourage public participation by knowledge and technology exchange, training and other forms of inclusive action (Art 6 UNFCCC, 1992).

The position of the UNFCCC and the Paris Agreement iterates the need for an all-inclusive governance as a means of adequately addressing the issue of climate change. The negotiations of the 26th Conference of Parties reiterated the need for collaborative climate adaptation across levels and between government and non-government players towards actualising the goals of the Paris Agreement and engender sustainable development and growth (Glasgow Climate Pact Decisions -/CMA.3 and -/CP.26).

The international framework on climate adaptation as laid out in UNFCCC, the Paris Agreement, Conference of Parties and other international instruments on climate change adaptation shows a detailed agenda; the challenge, however, is how effective these frameworks are in climate governance at the national and sub-national levels. The uniformity of a legal or regulatory framework on climate change adaptation will aid streamlined action targeted at identified vulnerabilities and other matters that may arise. A regulatory framework will also raise the level of compliance with internationally stipulated adaptation measures, especially in relation to sustainable development. Through such a framework, adaptation actions across sectors can be regulated appropriately and monitored (against a specific yardstick); thereby providing an avenue to know how much work has been done regarding research, policies, laws and their implementation. The requirement of a legal framework does not detract from the fact that adaptation has been taking place and has in fact been successful in some sectors and areas. However, effective adaptation measures in line with sustainable development will require a lot of coordinated and sequential action and financing across sectors, spheres of government, with international agencies and may often require regional and sub-regional activities and investments. Such a legal framework would give assurance of government's sincerity and transparency to investors and international actors in adaptation and developmental activities.

5 Climate Change Adaptation in South Africa

Climate change action in South Africa is geared towards securing the implementation of the UNFCCC regarding both climate change mitigation and adaptation (National Climate Change Response White Paper). South Africa's response to climate change has so far been based on policies. In resolving the challenge of adaptation to climate change, South Africa seeks to focus on the attainment of the sustainable development goals, especially the eradication of poverty and the promotion of development. This link between sustainable development and climate change adaptation resonates throughout the country's policies and plan of action on climate change (Kotzé et al., 2016; National Climate Change Response Green Paper, 2010; National Climate Change Response Policy White Paper, 2011; The National Climate Change Strategy, 2004).

5.1 *The National Policy on Climate Change*

The climate change policy response in South Africa began in 2004 with the 2004 National Climate Change Response Strategy. There were several documents which focused on mitigation and are therefore not within the purview of this chapter (The Long-Term Mitigation Scenario Study, 2004; The Vision, Strategic Direction and Framework for Climate Policy, 2008). The Green Paper titled "National Climate Change Response" (NCCR) of 2010, which documented the government's intentions and objectives on the subject matter of climate change was made available to the public for consultation and comments in early 2011 (For instance, the Green Paper was open for comments on a dedicated website). The comments and consultations on the Green Paper culminated in the National Climate Change Response Policy White Paper of 2011 (NCCRP). The White Paper is an executive policy (Fuo, 2013) and provides directions for climate adaptation within the country, having been adopted by the South African government as the document outlining the plan of action for climate change adaptation and mitigation.

Concerning adaptation, the White Paper aims to promote effective management of climate change impacts to promote social, economic and environmental well-being of the country. Also, the White Paper is based on long-term action towards a resilient and low carbon country (NNCRP White Paper). The long-term plan of action in relation to climate change here implies that adaptation, for instance, will not be limited to addressing present or apparent vulnerabilities, but will entail programmes or actions that span an extended period, the end aim being to attain resilience. The guiding principles and validation for the White Paper as a government policy flow primarily from the Constitution of the Republic of South Africa, particularly the Bill of Rights contained in Chapter 2, and the National Environmental Management Act at the national level. In international law, these validating principles are found in environmental law, human rights and sustainable development. These include principles such as the principle of common but differentiated responsibilities, equity, the needs and dignity of the poor and vulnerable, the principle of intra and inter-generational sustainability, the precautionary principle, the principle of informed public participation and so on (NCCRP, 2011). These set of principles emphasise the point that climate adaptation spans various frontiers of governance including decision-making, respect for human rights and the rule of law, infrastructural and economic development, as well as the involvement of all stakeholders in the adaptation process.

The White Paper, by highlighting sectors where the most action is required, confirms the country's obligations regarding the UNFCCC to, amongst other things, develop responses to climate change for coastal zones, water resources, land use, and integrate climate change into all social, economic and environmental policies (NCCRP, 2011). An important aspect of climate change adaptation and resilience also highlighted in the White paper is the building of capacity for disaster risk management as a means of addressing inevitable impacts of climate change. Given the reliance on sustainable development, the country's policy on climate change seeks to provide the most satisfactory response regarding available resources. Also, such responses aim at social and economic development (NCCRP, 2011). Bearing in mind that unemployment, poverty and informality are major challenges in South Africa, the approaches highlighted in the NCCRP ought to be adequately implemented to aid adaptation.

A characteristic of long-term climate change action is that its implementation is in phases or incrementally (Technical Paper on Long-Term Adaptation Planning, 2017). The climate change strategy outlined in the White Paper meets this criterion as it states that action on climate change entails incremental interventions. First, there is the implementation of climate change interventions already envisaged in existing legislation, regulations and policies, such as the Integrated Coastal Management Act (No. 24 of 2008) and the Disaster Management Act (No. 57). The next step would be the review of other laws, regulations and policies that have no climate change considerations to include benefits for climate change action in addition to the core provisions of such instruments. Finally, will be the fashioning of new and subsequent legislation to incorporate climate change considerations. In this regard, consultations are ongoing on the recently re-introduced Climate Change Bill of 2021. Further, in this regard, the government at the national level must audit existing policies and laws to ensure compliance not just with the White Paper, but with the Climate Change Act if eventually passed.

The White Paper also highlights the role of provincial governments in formulating strategies for climate change response. Also, the role of the local government as stipulated in the White Paper includes ensuring resilience through the planning of human settlements and urban development. While acknowledging the role of local governments, the White Paper adopts a rather tentative approach in highlighting the role of the local governments, for instance, it acknowledges the uncertainty in the role of local governments in taking up climate change action (NCCRP, 2011, p. 38) At this point, it is argued that the speculative tone employed in the policy document regarding the roles of local governments is a likely recipe for inaction or insufficient action in terms of climate change adaptation. Thus, to rectify this, it is argued that there would be a need for a binding legal framework setting out specific government obligations and the extent of each sphere's responsibility regarding climate change adaptation (NCCRP, 2011).

Another major component of adaptation action outlined in the White Paper is that it cuts across and engenders development in various sectors, including the water, agriculture, biodiversity and ecosystems, health, human settlements and disaster risk management sectors. Part of the aim of such spread is to protect and support vulnerable groups. Like most countries in Africa, South Africa, experience a high level of poverty and unemployment (Stats SA, Department of Statistics South Africa; World Bank, 2020), these factors have a tendency to exacerbate socio-economic vulnerability in South Africa (Ziervogel et al., 2014). It is thus necessary for these policy goals to be implemented to address socio-economic challenges such as access to energy, food, water and other basic necessities to raise the standards of living of the most vulnerable and thereby aid adaptation. It is noted here that the Department of Environmental Affairs at the national level of governance is presently working on a Climate Change Bill and a National Adaptation Strategy (NAS). The policy and legislation are aimed at synchronising climate change adaptation action in South Africa and to provide a common set of objectives for the mainstreaming of climate change into various sectors and institutional frameworks (Draft of the National Climate Change Adaptation Strategy, Republic of South Africa, 2017).

5.2 *Provincial Climate Change Policy*

The Western Cape Climate Change Response Strategy 2014 (WCCCRS) recognises the fact that traditional measures of addressing environmental and developmental challenges and disaster risk management have become insufficient and there is the need for improvement; particularly given climate change, its impacts and the need to adapt. The province aims to reduce vulnerabilities and increase resilience in the people, the economy, the ecosystem and infrastructure in line with its plans to actualise the sustainable development goals. In line with the national White Paper, the WCCCRS also employs an institutional framework that entails the government collaborating with non-governmental organisations, the private sector and businesses, academia and research community and other stakeholders. In this regard, as with the White Paper, the areas of implementation of climate change adaptation are cross-sectoral entailing various focus areas such as energy efficiency, renewable energy, infrastructure development, management of human settlements, transportation, management of biodiversity and ecosystem services, coastal management and food security and public health (Western Cape Government, 2014).

These interrelations in the adaptation process further buttress the expediency of legislation or regulatory framework to regulate the interactions between sectors and to lay down clear principles for climate change adaptation. The NCCAS extends the governance of climate action to and harmonises the goals of other related frameworks and policies linked to climate adaption such as the National Development Plan, the Sustainable Development Goals Agenda and the Sendai Framework on Disasters (NCCASS).

5.3 *The Local Government Climate Change Policy*

The City of Cape Town Framework for Adaptation to Climate Change identifies specific sectors which would most feel the impact of climate change and where the need for adaptation is greatest. It outlines a mode of operation to address identified vulnerabilities to climate change and prioritises the most suited adaptation measures. The Framework also highlights the need to follow the Integrated Development Plan. The IDP emphasises the need to address the impacts of climate change by improving on human development indicators and access to basic services, amongst others to increase economic, social development and eradicate poverty.

The City of Cape Town Climate Change Policy of 2017 (CCTCCP) builds on the Framework for Adaptation and other climate adaptation policies and initiatives. The City aims to mainstream climate change into all its policies and decision-making processes so that eventually there will no longer be a need for a stand-alone climate change policy (The City of Cape Town Climate Change Policy, CCTCCP, 2017). Regarding adaptation specifically, it aims to attain a "compact urban form" that is economically and socially inclusive. The CCTCCP also aims to take cognisance of climate change adaptation in the development of infrastructure for service delivery and to take a proactive approach to disaster risk management. This entails the management of risks in a way that increases social and economic development and the protection, preservation and management of natural systems to achieve a vibrant ecosystem and ensure the sustenance of ecosystem services (CCTCCP, 2017).

The principles guiding adaptation action in the CCTCCP aim at identifying impacts across sectors, improving resilience and reducing vulnerabilities, enabling a functional ecosystem that can provide necessary services and contribute to adaptation. The principles also promote an equitable spread of socio-economic development and encourage a behavioural change amongst members of the society (CCTCCP, 2017). To fulfil its adaptation objectives, the City set out to implement its policy collaboratively with other stakeholders, including other spheres of government, the private sector, the academic and research community, civil societies, non-governmental organisations and so on (CCTCCP, 2017).

A primary strategy the City seeks to employ in ensuring climate change adaptation is urban design and the development or maintenance of infrastructure. In a bid to achieve this, the City relies on urban planning and improvement of infrastructure, especially in areas where these are lacking, such as informal settlements. Given this, there is the need for incorporation of climate change considerations planning, land use and development control (CCTCCP, 2017), as well as an appropriate review of the relevant legislation to reflect the city's policy on climate change (CCTCCP, 2017).

In terms of the City's adaptation plan, it is noted that there have been conflicts between action required for climate change adaptation and environmental conservation and the need to foster socio-economic development within the City. A case in study is that of the Phillipi Horticultural Area (PHA) where the City granted applications for the rezoning of land within the area for the purpose of housing development. While it was acknowledged that the move was aimed at resolving the equally pressing and constitutionally recognised issue of housing, the decision of the City was argued to jeopardise food and water security, specifically because the development would threaten the Cape Flats aquifer within the area (Battersby-Lennard & Haysom, 2012). The conflict on the PHA was resolved by the Court in favour of the Phillipi Community. In its decision, the Court held that the City of Cape Town was obliged to give consideration to issues of water scarcity and climate change in reaching its decision on developments within the PHA (Philippi Horticultural Area Food & Farming Campaign and Another v MEC for Local Government, Environmental Affairs and Development Planning: Western Cape and Others, 2020).

The City's climate change policy has an implementation plan majorly aimed at incorporating climate change into the existing lines of action such as energy efficiency and access to energy, management of urban settlements and disaster risk management (CCTCCP, 2017). A practical approach to adaptation can be gleaned from the City's Integrated Development Plan which contains a strategic plan for building resilience and adapting to climate change. Areas of action identified under the IDP include flood control, water security, access to energy and other sectoral interventions necessary to manage the City's system and structures (City of Cape Town Integrated Development Plan). Further, the City of Cape Town Climate Change Action Plan of 2021 iterates the need for an integrated governance structure to address adaptation issues. This includes strategic focus areas and cross-cutting work across key areas and sectors including water, energy and economic development (City of Cape Town Climate Change Action Plan, 2021).

6 Climate Change Adaptation in Nigeria

Nigeria's response to climate change at the national level entails a national response strategy and a recently passed legislation at the national level which establishes a Climate Change Council to oversee and coordinate climate change action in the country. Prior to the Climate Change Act of 2021 (CCA), Nigeria's climate change response was based on policies. The policies on climate change are based on the Constitution of the country, the UNFCCC and its protocols and the principles of sustainable development.

In addition to the establishment of a Climate Change Council, the CCA also requires that climate change considerations are to be mainstreamed into government policies (Section 1(c) and (e) Climate Change Act, 2021). In terms of adaptation, specific action highlighted in the CCA are to identify risks and vulnerabilities, build resilience and strengthen adaptive capacity in actualising the Nation's long-term climate change adaptation plan (Section 1, Climate Change Act, 2021). The Act requires the Climate Change Council to formulate a Strategy, as well as a National Climate Change Action Plan which is to be reviewed every five years (Section 20 (1), Climate Change Act, 2021). In relation to adaptation, the Action Plan is to guide planning around climate change issues, establish national goals and objectives on climate adaptation (Section 20(4) Climate Change Act, 2021) and to identify specific areas of vulnerability.

It is noted however that the Act does not make provision for intergovernmental cooperation between the various levels of government regarding climate change action as the Act applies to only ministries, departments and agencies of the federal government (Sections 2 and 35 of the Climate Change Act, 2021). Further, section 25 of the Act which provides for collaboration and partnership on climate action, limits such action to partnership between the NCC, the ministry of environment and civil society organisations. In essence, the Act missed out on the opportunity to initiate and promote cooperation in climate action between all levels of government. Also, the Council on Climate Change is to be comprised of rather high-sounding officials—the Vice President of the Federation and Ministers of various Ministries and other politically appointed members of the executive arm of government (Section 5, CCA, 2021). It is here argued that the membership of the Council as provided for in the Act does not give room for sustainability and continuity of climate change governance and action in Nigeria.

6.1 National Policies on Climate Change

The National Strategy on Adaptation adopts the strategy of integrating its considerations and policy objectives into existing developmental programmes and projects. Thus primarily, climate change adaptation hinges on sustainable development (National Adaptation Strategy and Plan of Action on Climate Change for Nigeria [NASPA-CCN], 2011), as is the case in South Africa. Also, Nigeria's Vision 2020: Economic Transformation Blueprint highlights the need to act to address the impacts of climate change on socio-economic development in the country.

Responses to climate change adaptation in the national policy entail collaborative action between the three levels of government, the private sector and the civil society. The role of the federal government in the response strategy is mainly to provide policy and implementation guidance on climate change adaptation (NASPA-CCN, 2011). In reality, however, there is little guidance or collaboration between Lagos State and the federal government in these terms. On the contrary, Lagos State has been the trailblazer in the country regarding climate change action. According to the national policy, the states and local governments have the responsibility to initiate and carry out climate change adaptation within their various jurisdictions (NASPA-CCN, 2011).

It is noted that the policy does not provide guidelines or a framework for collaboration amongst the various stakeholders. This lack of structure on collaborative action is quite unlike the situation in South Africa where the White Paper provides a guideline for vertical and horizontal interaction and collaboration as well as cross-cutting and inter-sectoral climate change adaptation.

In the NASPA-CCN, the challenge of climate change is viewed as both an environmental (NASPA-CCN, 2011) and a developmental issue. Thus, the policy iterates that climate change adaptation should be integrated into the country's sustainable development Agenda, especially regarding economic development, provision and strengthening of infrastructure and food security and the management of the environment. The national policy also highlights the need for advocacy and public awareness on climate change, its impacts, the projections and the need for adaptation.

A significant adaptation strategy highlighted in the national policy is the use of indigenous knowledge and capacity, which implies a reliance on community-based adaptation (Learning from Experience – Community-based Adaptation to Climate Change). In promoting community-based adaptation, the national policy notes that resilience will be best achieved through a synergy between traditional and scientific knowledge. This implies that at all levels of government, steps should be taken to promote and build capacity in the use of traditional means of adaptation and also ensure that local communities benefit from the sustainable development agenda. That being said, research shows that in a series of projects carried out on community-based adaptation in Nigeria, it was discovered that major challenges encountered by communities in adaptation to climate change were the lack of access to land, insecurity of tenure and lack of infrastructure and basic services (Learning from Experience - Community-based Adaptation to Climate Change).

To build capacity for action on climate change adaptation, the national policy pays attention to the plight of women, children and other people vulnerable groups. This is a major point because, in most developing countries like Nigeria, women and children bear the brunt of social and economic instability of lack of resources. The national policy identifies specific sectors where the impacts of climate change will be most felt and requires adaptation action to begin from such sectors. These include water, agriculture and food, biodiversity and ecosystems, coastal zones, human settlements, disaster risk management and so on. The fact that climate change impacts are all-encompassing means that action in that regard also must be multi-dimensional, flexible and continuous (Frohlich & Knieling, 2013). Thus, a major part of the action plan is to enact legislation on climate change, and mainstream climate change adaptation into the National Development Plans across government departments and agencies (NASPA-CCN, 2011). The policy also requires States to take similar steps, and local governments are mainly required to provide the capacity for community-based adaptation (NASPA-CCN, 2011).

In addition to legislation and instruments already highlighted, the national policy is guided by the United Nations Development Group principles which are based on a human rights approach, gender equality, environmental sustainability and capacity building (NASPA-CCN, 2011). These principles, according to the national policy, are linked to the fundamental objective and directive principles contained in Chapter 2 of the Constitution for validity and implementation. The unsettling issue here is that Chapter 2 of the Constitution is by the Constitution itself rendered non-justiciable and unenforceable. This non-justiciability casts doubts as to the enforceability of the climate change response strategy itself. This point is further addressed later in this chapter.

6.2 *Climate Change Action at the State Level*

Lagos State has by far been the most proactive regarding action on climate change in Nigeria (Lagos State Development Plan 2012–2025) (Towards a Lagos State Climate Change Adaptation Strategy), yet it has no stand-alone policy on climate change adaptation. The state addresses issues of climate change through its development plan. Action taken by the state to address climate change include advocacy and public awareness, developmental activities such as the improvement of basic services and infrastructure and increased reliance on physical planning strategies. Lagos State, having held Summits on climate change annually since 2008 commissioned a project on a suitable climate change strategy, this culminated in the Report, "Towards a Lagos State Climate Change Adaptation

Strategy (LAS-CCAS) 2012". The Report, which is a precursor to the Lagos State policy on climate change, aligns with the guidelines set by the national policy.

The LAS-CCAS focuses majorly on adaptation to climate change and the approach outlined in the Report cuts across the same sectors highlighted in the national policy. The Report is not a policy document. Thus, it mainly describes steps and responses that Lagos State has employed in tackling the issue. The LAS-CCAS also makes recommendationswhich include the need for a policy on climate change, proper planning for climate change adaptation action and improved capacity on information management (this is particularly essential for disaster risk management). It also makes recommendations on risk reduction and livelihood development. The recommendations also highlight the need for gender considerations, especially the need to build capacity and encourage individual adaptation to the challenges of climate change.

The LAS-CCAS requires that climate change action should be long-term and should include the relocation of communities along the Lagos coast and strict control for land reclamation along the state's coast. This chapter highlights these because the state is already conflicting with these recommendations with the building of the Eko Atlantic City and the eviction of the residents of informal settlements such as the Otodo-gbame community within the State. It is here argued that adaptation initiatives of the government may end up leading to maladaptation if they do not take into consideration environmental impact and human rights of vulnerable groups, for instance, the land reclamation exercises being carried out to accommodate the new Eko Atlantic City and other new exclusive settlements on the Islands of Lagos could end up resulting in maladaptation, especially where the environmental assessments for the reclamation did not take into consideration the likely adverse effects of such an activity on other areas of Lagos State.

The Lagos State Development Plan 2012–2025 (LSDP) integrates climate change considerations as a significant factor towards achieving sustainable development. The LSDP recognises that adverse impacts of climate change are a bane to social, economic and environmental well-being; thus, it considers climate change as a factor in risk management and safety, environmental management and protection, protection of ecosystem and biodiversity, urban development and sustainable development (LSDP 2012–2025).

7 Assesssment of Disaster Risk Management in the Adaptation Process

The research evaluates disaster risk management as an essential aspect of the adaptation process in both countries. This is based on the major link between weather and disasters, and many communities in both Nigeria and South Africa are prone to these weather-related disasters, especially since the changes in climate aggravate these disasters (Lei & Wang, 2014). It is noted that climate change could destabilise pre-existing social and environmental resilience, thereby leading to new vulnerabilities (Lei & Wang, 2014). The UNFCCC notes that climate change comes with adverse effects (The Preamble to the UNFCCC, 1992), which are damaging and harmful to the ecosystem, human health and welfare and socio-economic conditions within the society (Art 1 UNFCCC, 1992). The Cancun Framework thus requires member states to take action on disaster risk management and take into consideration the Hyogo Framework for Action regarding early warning signs, risk assessment and management. The Paris Agreement iterates this need for disaster risk management (Art 7(5) of the Paris Agreement).

The Hyogo Framework requires that states take steps to increase public awareness of the disaster risks associated with climate change across various sectors (International Strategy for Disaster Reduction, Hyogo Framework for Action 2005–2015, 2005–2015). The Hyogo Framework also states the need to incorporate climate risks into future measures to address and reduce the risk of disasters as a result of climate change (para 19 (c), para 30 Hyogo Framework for Action 2005–2015). The Sendai Framework for Disaster Risk Reduction 2015–2030, which is the successor Framework to the Hyogo Framework, notes that between 2008 and 2012 about 144 million people were displaced by disasters, which were made worse by climate change. The Sendai Framework also notes the need to reduce vulnerabilities to climate change so that that would, in turn, reduce the disaster risks associated with it (Paragraph 6 and 10 Sendai Framework for Disaster Risk Reduction 2015–2030).

The Sendai Framework in its Guidelines also iterates the need to integrate disaster risk reduction into policies and plans relating to climate change and sustainable development, amongst other factors, thereby establishing an irrefutable link between climate change, disaster risk reduction and sustainable development (Paragraph 19 Sendai Framework). In recognition of the continuous and incremental action on climate change, the Sendai Framework acknowledges that disaster risk reduction must be proactive, in that there should be a frequent review and update of disaster risk reduction plans and measures (Paragraph 33, Sendai Framework).

While disaster risk management features in the adaptation policies of both case studies, Nigeria has no legal framework on disaster risk management. In South Africa, disaster risk management is regulated by the Disaster Risk Management Act No 57 2002 (DRMA).The DRMA provides for the "integrated and co-ordinated" management of disasters and stipulates that a DRM plan must provide for an integral part of planning at the province and municipality (sections 38(2), 39(2) (*a*), 47 (1) (*c*), and 53(2) (*a*) DRMA). To this end, the Act focuses on establishing a DRM framework at all spheres of government for the purposes of prevention or reducing of disaster risks, mitigating the seriousness of disasters and preparedness and responding to such disasters that might occur (sections 7, 38 DRMA). The aim of the framework is to reduce vulnerability in places, communities and households found to be disaster-prone (section 7(2) DRMA).

Both countries considered in the chapter address climate change adaptation through policies at all levels or spheres of government. Specific tools for adaptation include inter-sectoral and inter-governmental collaboration, as well as public participation by international, regional, national, local and private stakeholders. While South Africa's NCCRP White Paper provides a detailed framework for the various collaborative action, Nigeria's NASPA-CCN provides no framework for such collaboration or inter-relationship. The chapter highlights this as a significant shortcoming in climate change action in Nigeria. Further, while significant climate action takes place at the municipal level in South Africa, the local government level which is the third level of governance in Nigeria plays little or no roles in climate change adaptation governance. This is based on the general relegation of local governments in Nigeria (Adeyemi et al., 2012; Olaiya, 2016), while the activities of municipalities in South Africa are hampered by various factors such as funding and technical capacity (du Plessis & Kotze, 2014).

Further, climate change adaptation policies in both South Africa and Nigeria note that disaster risk management is an essential sector for action on climate change adaptation. The approach to disaster risk reduction in South Africa, especially in the Western Cape, is proactive and is on a short-term basis aimed at reducing vulnerability to the adverse effects of climate change (van der Berg et al., 2016). On the other hand, the outlook for disaster management in Lagos State requires a lot more action across levels in order to sufficiently address issues of adaptation to climate change. For instance, since the historic flooding of Lagos in June 2011 and 2012 (IFRC, 2011; Vangaurd, 2012), Lagos has experienced almost annual flood, yet research shows that the State has not been able to provide flood disaster risk management address the recurring flooding risk within the city-state (Adelekan, 2016). A key objective of disaster risk reduction in both countries is to reduce vulnerabilities in human systems (that is within communities where people live), and in natural systems (the marine, terrestrial, freshwater ecosystem) (van der Berg et al., 2016). The reduction of vulnerabilities in human systems implies the provision of basic infrastructure and services. In relation to natural systems, it implies the protection and conservation of natural environments that act as buffers and absorbers of natural occurrences that would otherwise result in disasters. Here again, an undeniable link is recognised between climate change adaptation, disaster risk reduction and development. It is noted that disaster risk management has a major role to play in adaptation to climate change. For instance, the establishment of new coastal management lines in the City of Cape Town is noted as a proactive step towards disaster management.

Given the discussion above, adaptation framework, which integrates robust disaster risk management must entail an assessment of the new risks arising or likely to arise, and preparedness to respond and recover appropriately (Solecki et al., 2011). This point buttresses the application of the precautionary principle of environmental law to adaptation action. The absence of sufficient information should not be a reason not to take necessary action on climate change and associated risks. It is also opined that disaster risk reduction and management should be an integral aspect of community-based adaptation. This is because most disasters occur at the local or community level (Lei & Wang, 2014; van der Berg et al., 2016), and adaptation governance at that level ought to be capable of addressing it adequately as first responders. Also, disaster risk reduction, as an aspect of adaptation to climate change, should be community specific, taking into cognisance the specific vulnerabilities or hazards communities are prone to and how they can be addressed.

8 Conclusion

This chapter analysed and assessed the existing framework on climate change adaptation in both Nigeria and South Africa by way of desktop analysis. Both countries have addressed climate change adaptation through policies at all levels or spheres of government. The analysis shows the policies of both countries focus on reducing vulnerabilities and increasing resilience to adequately adapt to the adverse effects of climate change. Specific tools for adaptation include inter-sectoral and inter-governmental collaboration, as well as public participation by international, regional, national, local and private stakeholders. While South Africa's NCCRP White Paper provides a detailed framework for the various collaborative action, Nigeria's NASPA-CCN provides no framework for such collaboration or inter-relationship. The thesis highlights this as a significant shortcoming in climate change action in Nigeria.

The policies of both countries require that climate change considerations should be mainstreamed into the legislation and policies of the most affected sectors but have failed to provide a strategy or guide for such implementation. There is also yet no mechanism to monitor the level of compliance with the national policies on climate change, especially regarding adaptation. This is a shortcoming, and this thesis presents this as a basis for a legal framework for climate change adaptation. Sustainable development features strongly in the policies of both countries as an underlying theme of climate change adaptation. It is a fundamental aspect of two major approaches to climate change adaptation highlighted in the policy documents of both countries, that is, community-based adaptation and ecosystem-based adaptation. For adaptation to be effective, it must be based on social and economic development, equity and environmental protection. The analysis in this chapter points to the fact that there is ongoing action on adaptation to climate change, but a lot still needs to be done. This is particularly the case in Lagos, Nigeria, where there is a considerable knowledge gap amongst the populace. In South Africa, the major challenge is that of implementation of the policies and the fact that the municipalities are overburdened; without the assistance of the provincial and national governments, adaptation to climate change would be an added burden.

References

Adelekan, I. O. (2016). Flood risk management in the coastal city of Lagos, Nigeria. *Journal of Flood Risk Management*, 255–264.

Adeyemi, O. O., Akindele, T. S., & Aluko, O. A. (2012). Institutionalizing the culture of accountability in local government administration in Nigeria. *African Journal of Political Science and International Relations, 6*(4), 81–91. https://doi.org/10.5897/AJPSIR11.127

Battersby-Lennard, J., & Haysom, G. (2012). *Philippi horticultural area: A city asset or potential development node?*

Bauer, A. (2014). Multi-level governance of climate change adaptation through regional partnerships in Canada and England. *Geoforum*, 121–129.

Bridge, G., & Perreault, T. (2009). Environmental governance. In N. Castree, D. Demeritt, D. Liverman, & B. Rhoads (Eds.), *A guide to environmental geography* (pp. 475–492). Wiley-Blackwell.

Chaffin, B. C., Gosnell, H., & Cosens, B. A. (2014). A decade of adaptive governance scholarship: Synthesis and future directions. *Ecology and Society, 19*(3).

City of Cape Town. (2021). *City of Cape Town Climate Change Action Plan*. City of Cape Town.

City of Cape Town. (2017 July–2022). *City of Cape Town Integrated Development Plan*. City of Cape Town.

Climate Change Act. (2021).

Corfee-Morlot, J., Kamal-Chaoui, L., Donovan, M. G., Cochran, I. R., & Teasdale, P. J. (2009). *Cities, climate change and multilevel governance* (OECD Environmental Working Papers, 24).

Dellmuth, L. M., & M, G. (2021). Global adaptation governance: How intergovernmental organizations mainstream climate change adaptation. *Climate Policy*, 868–883.

Draft of the National Climate Change Adaptation Strategy, Republic of South Africa. (2017). Department of Environmental Affairs.

du Plessis, A., & Kotze, L. (2014). The heat is on: Local government and climate governance in South Africa. *Journal of Africa Law, 58*(1), 145–174.

Frohlich, J., & Knieling, J. (2013). Conceptualising climate change governance. In J. Knieling, & W. Filho (Eds.), *Climate change governance* (p. 20).

Fuo, O. 'Constitutional basis for the enforcement of "executive" policies that give effect to socio-economic rights in South Africa' (2013) (16) 4PER / PELJ. PER/PELJ, 4(16).

Glasgow Climate Pact Decisions -/CMA.3 and -/CP.26.

Hatfield-Dodds, S., Nelson, R., & Cook, D. C. (2007). *Adaptive governance: An introduction and implications for public policy* (No. 418–2016-26492).

Hooghe, L., & Marks, G. (2001). Types of multi-level governance. *European Integration online Papers (EIoP), 5*(11).

Huitema, D., Adger, W. N., Berkhout, F., Massey, E., Mazmanian, D., Munaretto, S., Plummer, R., & Termeer, C. C. (2016). The governance of adaptation: choices, reasons, and effects: Introduction to the special feature 21(3) 37 at 45. *Ecology and Society*, 37.

Hyogo Framework for Action 2005–2015.

IFRC. (2011). *Nigeria: Flash floods*. Retrieved from www.ifrc.org/docs/appeals/11/MDRNG010.pdf.

International Strategy for Disaster Reduction, Hyogo Framework for Action 2005–2015. Retrieved from https://www.unisdr.org/we/coordinate/hfa

Ishtiaque, A. (2021). Multilevel governance in climate change adaptation: Conceptual clarification and future outlook. In F. Ali (Ed.), *Climate change and extreme events* (pp. 171–185). Elsevier.

Janicke, M., & Jorgens, H. (2006). New approaches to environmental governance. In M. Jänicke & K. Jacob (Eds.), *Environmental governance in global perspective: New approaches to ecological modernisation* (pp. 167–202). Freie Universität Berlin.

Jordan, A., & Huitema, D. (2014a). Innovations in climate policy: Conclusions and new directions. *Environmental Politics*, 906–925.

Jordan, A., & Huitema, D. (2014b). Innovations in climate policy: The politics of invention, diffusion, and evaluation. *Environmental Politics*, 715–734.

Jordan, A., & Huitema, D. (2014c). Policy innovation in a changing climate: Sources, patterns and effects. *Global Environmental Change*, 387–394.

Kotzé, L., Humby, T., Rumble, O., Gilder, A., & Lehmann, K. (2016). Climate change law and governance in South Africa- Setting the scene. In T. Humby, L. J. Kotzé, O. Rumble, & A. Gilder (Eds.), *Climate Change: Law and Governance in South Africa* (pp. 1–19).

Lagos State Development Plan 2012–2025.

Learning from Experience—Community-based Adaptation to Climate Change. (n.d.). Retrieved from www.nestinteractive.org/publications.php?OrderPublication

Learning from Experience – Community-based Adaptation to Climate Change. (n.d.). pp. 12–13. Retrieved from www.nestinteractive.org/publications.php?OrderPublication.

Lei, Y., & Wang, J. (2014). A preliminary discussion on the opportunities and challenges of linking climate change adaptation with disaster risk reduction. *Natural Hazards*, 1587–1588.

Madzwamuse, M. (2011). *Climate governance in Africa- adaptation strategies and institutions*.

Massey, E., Biesbroek, G. R., Huitema, D., & Jordan, A. (2014). *Climate policy innovation: The adoption and diffusion of adaptation policies across Europe*. Global Environmental Change.

MISTRA Urban Futures. (2015). *Understanding climate adaptation and transformation challenges in African cities* (Mistra Urban Future Policy Brief 2015:1).

Nash, S. L., Torney, D., & Matti, S. (2021). Climate Change acts: Origins, dynamics, and consequences. *Climate Policy*, 1111–1119.

NASPA-CCN. (2011). National Adaptation Strategy and Plan of Action on Climate Change for Nigeria (NASPA-CCN).

National Climate Change Response Green Paper (2010).

National Climate Change Response Policy White Paper (2011).

Olaiya, T. A. (2016). Governance and constitutional issues in the Nigerian local governments. *Journal of Politics and Law, 9*(1), 21–47.

Patterson, J. J., & Huitema, D. (2019). Institutional innovation in urban governance: The case of climate change adaptation. *Journal of Environmental Planning and Management, 62*(3), 374–398.

Philippi Horticultural Area Food & Farming Campaign and Another v MEC for Local Government, Environmental Affairs and Development Planning: Western Cape and Others, [2020] ZAWCHC 8 (Western Cape High Court 2020).

Schultz, L., Folke, C., Österblom, H., & Olsson, P. (2015). Adaptive governance, ecosystem management, and natural capital. *Proceedings of the National Academy of Sciences, 112*(24), 7369–7374.

Sendai Framework for Disaster Risk Reduction 2015–2030. Retrieved from http://www.unisdr.org/we/inform/publications/43291

Solecki, W., Leichenko, R., & O'Brien, K. (2011). Climate change adaptation strategies and disaster risk reduction in cities: connections, contentions, and synergies. *Current Opinion in Environmental Sustainability*, 136.

Stats SA, Department of Statistics South Africa. (). Poverty in Perspective.

Stern, N. (2007). *The economics of climate change: The Stern review*. Cambridge University Press.

Taylor, A. (2016). Urban adaptation. In T. Humby et al. (Eds.), *Climate change: Law and governance in South Africa* (pp. 11–12).

Technical Paper on long-term adaptation planning. (2017). *Outcome of the 12th meeting of adaptation committee* (p. 6). Retrieved from www.unfccc.int/files/adaptation/groups-committee/application/pdf/ac12_7b-longterm.pdf

The City of Cape Town Climate Change Policy. (2017). CCTCCP. Retrieved from www.capetown.gov.za/Document-centre

The Long-Term Mitigation Scenario Study. (2004).

The National Climate Change Strategy. (2004).

The Paris Agreement. (2015).

The Vision, Strategic Direction and Framework for Climate Policy. (2008).

Towards a Lagos State Climate Change Adaptation Strategy. (n.d.).

UNDP Climate Promise. (2022, August). *In the race against climate change, there's still hope for Africa—Here are 5 reasons why*. Retrieved from UNDP: https://climatepromise.undp.org/news-and-stories/race-against-climate-change-theres-still-hope-africa-here-are-5-reasons-why

United Nations Framework Convention Framework on Climate Change. (1992). United Nations Framework Convention Framework on Climate Change.

van der Berg, A., Plessis, A. D., & Murphree, M. (2016). Local disaster risk reduction and management in South Africa's response to climate change. In T. Humbly et al. (Eds.), *Climate Change Law and Governance in South Africa* (pp. 12–15). Juta

Vangaurd. (2012, June). *7 feared dead as flood sweeps Lagos*. https://www.vanguardngr.com/2012/06/7-feared-dead-as-flood-sweeps-lagos/

Western Cape Government. Western Cape Climate Change Response Implementation Framework (2014).

World Bank. (2020). *Poverty and equity brief: South Africa*.

Ziervogel, G., New, M., Archer van Garderen, E., Midgley, G., Taylor, A., Hamann, R., Stuart-Hill, S., Myers, J., & Warburton, M. (2014). Climate change impacts and adaptation in South Africa. *WIREs Clim Change*, 605–620.

Zurn, M. (2010). Global governance as multi-level governance. In M. Zurn, S. Walti, & H. Enderlein (Eds.), *Handbook on multilevel governance*. Edward Elgar Publishing, Inc.

Zürn, M., Wälti, S., & Enderlein, H. (2010). Introduction. In H. Enderlein, S. Wälti, & M. Zürn (Eds.), *Handbook on multilevel governance*. Edward Elgar Publishing, Inc.

CHAPTER 7

Legal Responses to Climate Disasters in Southern Africa: Practical Legal Developments in Zimbabwe and Mozambique

Lenin Tinashe Chisaira

1 Introduction

Southern Africa is a sub-Saharan region comprising fifteen States that share similar characteristics with other least developed to low-income countries, namely inequality, inadequate resources, environmental and climate injustice poor development and lack of effective implementation of progressive laws and policies (Zungu et al., 2021). Climate justice refers to a political-economic-legal framework that brings together "the intersection between climate change and the way social inequalities are experienced as structural violence" (Godden et al., 2022).

L. T. Chisaira (✉)
Advocates for Earth, Harare, Zimbabwe
e-mail: tinashe@advocates4earth.org

M. Addaney et al. (eds.), *Climate Change in Africa*,
https://doi.org/10.1007/978-3-031-30050-9_7

From the 2000s onwards, two Southern African states, namely Zimbabwe and Mozambique, have experienced frequent climatic hazards with several implications for social equality. These implications include legal impacts such as expectations of human and environmental rights protection by the State in the time of natural hazards and expectation of compensation for loss and damage occasioned by climate-induced natural hazards. In addition legal issues extent to environmental injustice, the right to food security, threats to the right to life, water, peace, development, shelter, health and gender equality (Nyahunda et al., 2021). Human rights impacts (Fershee, 2009) of climate hazards are worth discussing since these are naturally worsened in the time of a disaster. It is pertinent to make a comparative analysis of the legal and policy responses to climate change from these two case-study States, using climate justice and critical legal theoretical lenses.

2 The Context of Human Rights and Climate Hazards in Southern Africa

Climate hazards pose challenges to human rights (Atapattu, 2009). This is in addition to natural discomforts occasioned by unexpected weather and ecological and climatic changes (Hayes et al., 2018). Governments have drafted public policies to deal with climatic hazards. Zimbabwe's National Climate Change Response Strategy defines a climatic hazard in the following terms:

> Climatic hazard is any event or change in climate, such as a single extreme event that exceeds a critical temperature threshold or a complex combination of changes involving variables and/or resulting in multiple impacts. It is an extreme climatic/weather event causing harm and damage to people, property, infrastructure and land-uses. It includes not only the direct impacts of the climate/weather event itself by also other indirect hazards triggered by that event. (Government of Zimbabwe, National Climate Change Response Strategy, 2014, p. xx)

These defined climatic hazards have been increasing in frequency in the Global South, especially across South East Asia, Latin America and Southern Africa (Berbés-Blázquez et al., 2017; IPCC, 2022). In the twenty-two-year period between 2000 and 2022, Zimbabwe and Mozambique suffered from several cyclones, tropical storms, drought, diseases and heatwaves. Some of the main tropical cyclones to hit the two countries in the aforementioned period include Cyclone Idai, Cyclone Eline, Cyclone Fantala, Cyclone Dineo, Cyclone Kenneth and Cyclone Batsirai (Nhamo & Chikodzi, 2021). In addition, there were Tropical Storms Desmond and Eloise and severe droughts. These hazards resulted in massive loss of lives, for instance, Cyclone Idai alone killed hundreds of people in countries like Zimbabwe, Mozambique, Malawi and Madagascar (Nhamo & Chikodzi, 2021). In addition, over US$2 billion was calculated as the amount needed to cover the damage and to enable lives and infrastructure to return to normal following the cyclone (World Bank, 2019).

States institute a variety of legal, fiscal and policy responses to mitigate the adverse effects of climatic hazards. State responses range from statutory enactments, climate funds mobilisation and policy pronouncements to ratification and implementation of treaty decisions adopted by regional and intergovernmental bodies such as the United Nations (UN), the African Union (AU) and the Southern African Development Community (SADC) (Maluwa, 2012). Climate laws mainly deal with regulating mitigation and adaptation (Bodansky et al., 2017). Furthermore, Climate Acts are the legal avenue to provide for the setting up of climate funds that is fiscal resources set aside to deal with climate adaptation and mitigation. This is the case with the Climate Change Act of 2008 in the United Kingdom and the Climate Change Act of 2016 in the Republic of Kenya.

Legal responses to environmental and climate hazards have also emanated from advocacy lobbying by civil society and environmental groups (Percival & Zhao, 2014).In the SADC region, some of the active environmental groups that have taken some action concerning climate disasters include Friends of the Earth, La Via Campesina, Advocates4Earth, Natural Justice and the Centre for Environmental Rights.

The development of climate laws, policies and climate litigation in the Global South is one of the consequences of the impact of the climate crisis on the Global South's vulnerable population groups, businesses, governance operations and funding needs (Setzer & Benjamin, 2020). The climate crisis has arguably become an existential threat in general terms and specifically in terms of legal obligations and human rights and environmental justice expectations (IPCCC, 2022). Indigenous peoples in Africa and Latin America face perennial socio-ecological challenges including drought, displacements, exclusion, information, shortage, diseases and loss of life, property and livelihoods in the event of climate-induced hazards (Berbés-Blázquez et al., 2017). Such a crisis naturally requires policy innovations by governments and other stakeholders (Jordan & Huitema, 2014a). The climate crisis and its states of vulnerability require resilience building, adequate legal and policy responses, financing and economic regulation to uplift Global South individual and community well-being in the Anthropocene era.

Developing States and their clash with climate change have a public interest incentive to develop human rights laws that tally with internationally developed norms, obviously taking cognisance of cultural relativism. International human rights law does provide a cushion to protect victims (Bellinkx et al., 2022; Chowdhury & Bhuiyan, 2010). The Universal Declaration of Human Rights (UDHR), for instance, is an example of a soft law but is a highly authoritative declaration that has attained the status of customary international law. The UDHR provides that:

> Everyone has the right to a standard of living adequate for the health and well-being of himself and of his family, including food, clothing, housing and medical care and necessary social services. (UDHR, Article 25(1)).

Analysing the steps towards practical legal actions by affected states is necessary. Such analyses can save as a baseline to initiate meeting the obligations of industrialised economies to assist in meeting climate justice and financing.

3 Responses to Climate Change: A Political Economy and Legal Analysis

Globally in the modern era of industrial capitalism, for regions that already experience human, political and legal rights violations, the increasing impacts of climate injustice only serve to exacerbate vulnerability (Koch, 2011). While this chapter focuses on responses by governmental bodies, it is prudent to mention at the beginning that the initiative for policy and law reform can emanate from several sectors beyond government. Jordan and Huitema (2014b) outline that:

> ...politicians are definitely not the only actors involved in policy invention. In fact, their role comes relatively late in the invention process, whereas earlier stages are dominated by actors from outside the state (businesses,

academics, NGOs, international organizations) and by the national bureaucracy (the other key actor within the state). (Jordan & Huitema, 2014b, p. 390)

However, the focus on state and government response is influenced by the heavy role of the state in Southern Africa where democracy had not yet taken efficient roots. In this case, disasters occur when States and responsible authorities fail to cushion vulnerable population groups and communities from hazards and humanitarian crises (Cubie, 2019), including from the impacts of extreme weather events such as tropical cyclones or other events. Climate injustice combined with weak institutions and budgetary constraints has led to the current tragic human rights violations in Zimbabwe and Mozambique during natural hazards.

The SADC Meteorological Centre publishes reports on the impacts of the climate crisis within the region (Bopape et al., 2019). The latest cyclone and tropical storm seasons have led to the displacement of rural and coastal peoples, destruction of property, livelihoods, and futures, disappearances, sexual violence in displacement camps, as well as disruptions to education, roads and health services. Generally, "the ecological effects of climate change on societies and polities in many developing countries, already plagued by chronic economic, financial and political fragility, could be incalculable" (Abouyoub, 2012, p. 170). Climate hazards in the SADC region, as in anywhere else, turn into disasters when interspersed with existent economic and political crises. Climate-induced disasters serve to expose the ongoing hardships encountered by vulnerable population groups.

The United Nations has come up with several multilateral agreements and treaties providing for climatic hazards and the protection of human rights. The United Nations Framework Convention on Climate Change (UNFCCC) is the primary convention that deals with climate change. The climate treaty was adopted on 09 May 1992 and entered into force on 21 March 1994. Mozambique and Zimbabwe are both non-Annex 1 parties to the UNFCCC. Non-Annex I parties are essentially developing countries. Zimbabwe ratified the UNFCCC on 3 November 1992 while Mozambique ratified the same convention on 25 August 1995.

It is essential to ensure that climate-vulnerable countries take steps in enacting climate legislation that alleviates the suffering of their vulnerable sectors from climate change with support from Annex II countries as mandated by the UNFCCC. The main objective of the UNFCCC is to achieve the "stabilization of greenhouse gas concentrations in the atmosphere at a level that would prevent dangerous anthropogenic interference with the climate system" (UNFCCC, art 2). The convention indicates that developed countries various forms of support such as funding, insurance and technology transfer to meet the specific needs and concerns of among others, "'countries' with areas prone to natural disasters" (UNFCCC, art 4(8) (d)). More importantly, the convention outlines the need to ensure that legal responses are enacted by states in line with their varying contexts. In this regard, the convention recognises that "that States should enact effective environmental legislation" and other measures or standards that "should reflect the environmental and developmental context to which they apply, and that standards applied by some countries may be inappropriate and of unwarranted economic and social cost to other countries, in particular, developing countries" (UNFCCC, preamble). Zimbabwe and Mozambique should take these steps as matters of urgency.

As a framework convention, the UNFCCC is strengthened by additional protocols, agreements and declarations, one of these being the Paris Agreement. The Paris Agreement is a multilateral environmental agreement negotiated by parties to the UNFCCC at CoP21 in France. The agreement was adopted on 12 December 2015 and entered into force on 4 November 2016. The Paris Agreement outlines and invites some crucial considerations on climate change and mitigation. The Paris Agreement recognises that there are "specific needs and special circumstances of developing country parties" and that States "may be affected not only by climate change but also by the impacts of the measures taken in response to it" (Paris Agreement, Preamble). The Paris Agreement covers aspects of climate change mitigation, adaptation and climate finance. The agreement is critical in recognising the importance of human rights in the climate change discourse. The most important provision relating to human rights is an acknowledgement in the preamble of the agreement:

> Acknowledging that climate change is a common concern of humankind, Parties should, when taking action to address climate change, respect, promote and consider their respective obligations on human rights, the right to health, the rights of indigenous peoples, local communities, migrants, children, persons with disabilities and people in vulnerable situations and the right to development, as well as gender equality, empowerment of women and intergenerational equity. (Paris Agreement, Preamble)

The Paris Agreement is an essential cog in the link between human rights and climate change. Adelman & Lewis indicate that the agreement is "the first multilateral climate change instrument that refers to human rights" (Adelman & Lewis, 2018, p. 11). Adelman and Lewis also argue that "[T]he inclusion of a single reference (to human rights) in the Preamble rather than the operative part of the Agreement is less than human rights advocates had hoped for, but nonetheless constitutes an advance". However, while some view the token direct reference to human rights in the Paris Agreement as inadequate, the agreement is still a progressive step in the right direction.

The relevance of climate hazards on people's rights is buttressed by their emphasis in several other multilateral agreements. The United Nations Declaration on the Rights of Peasants and other People Working in Rural Areas (UNDRIP) emphasises that climate knowledge is one of the key rights for peasants. The declaration provides that:

> Peasants and other people working in rural areas have the right to adequate training suited to the specific agroecological, sociocultural and economic environments in which they find themselves. Issues covered by training programmes should include, but not be limited to, improving productivity, marketing, and the ability to cope with pests, pathogens, system shocks, the effects of chemicals, climate change and weather-related events. (UNDRIP, art 25 (1)).

There is a need to utilise additional multilateral agreements to ensure adequate policy planning by the Global South States. One of the agreements is the Sendai Framework on Disaster Management. Article 38 of the Sendai Framework The framework emphasises that "developing countries require an enhanced provision of means of implementation, including adequate, sustainable and timely resources, through international cooperation and global partnerships for development, and continued international support, so as to strengthen their efforts to reduce disaster risk"(Sendai Framework, Article 38). Multilateral agreements are also critical in ensuring human dignity and global development. Climatic hazards pose challenges even to development and world peace. The African Strategy on Climate Change enunciates this view in the following terms: "Climate change is bringing additional burdens to existing challenges such as displacement of people, internal and international migration and conflict" (Draft Africa Climate Change Strategy 2020–2030, p. 27).

Human beings, due to the universal and inalienable nature of human rights, deserve the peace of mind and the same adequate protection of their rights regardless of location or class. In the era of climate change, conflict and climate injustice, vulnerable population groups in countries like Zimbabwe and Mozambique usually end up with more violated rights under a blanket of impunity. The international human rights framework dictates that human rights be universal and inalienable. However, on the ground, the most critical aspect of human rights protection is vitiated by several environmental constraints, including budgetary and political will, in the context of this chapter, the sub-Saharan African States of which Zimbabwe and Mozambique are a part. In this regard, Donnelly argues that:

> The global human rights regime relies on national implementation of internationally recognized human rights. Norm creation has been internationalized. Enforcement of authoritative international human rights norms, however, is left almost entirely to sovereign states. (Donnelly, 2007, p. 283)

Human rights protection and laws in Southern African states such as Zimbabwe and Mozambique should hence be responsive to the demands of an era of climatic hazards.

3.1 Regional and Sub-Regional Policy Responses to Climate Impacts.

Regional institutions such as the AU and SADC play an integral role in shaping legal developments at the local and international levels. Regional bodies are critical in coming up with practical and sometimes uniform but region-specific developments that shape member states' response to crises. Schulz defines regionalism as "the body of ideas, values and concrete objectives that are aimed at creating, maintaining or modifying the provision of security and wealth, peace, and development within a region" (Schulz, 2001, p. 5; see also, Söderbaum, 2013). Regional institutions are expected to serve as accessible and vigilant platforms for monitoring climate change and human rights violations. This is so, because of the expectations of shared understandings within similar geographical settings.

Regional governance bodies are usually influential to the nation-state (Panke & Stapel, 2022). These bodies have a pivotal role to play in monitoring and placing a check on the impacts of structural inequalities imposed by climate injustice. From the top, the Charter of the United Nations recognises the importance of regional bodies (UN Charter, art 52 (1)). The increased resilience of the peoples of vulnerable states such as Mozambique and Zimbabwe will go a long way in lessening and even eradicating climate injustice-related conflicts and human rights violations. SADC has sought to have a unified voice during climate change negotiations as a body and in liaison with other African States in general under the African Group of Negotiators. SADC common regional responses to climate change include the SADC Policy Paper on Climate Change, the SADC Water Sector Climate Change Adaptation Strategy and the establishment of the SADC Climate Services Centre (Pegasys, 2014). Regions that are vulnerable to climate change-related hazards should have the capacity and the mechanisms for lessening the impacts of climate change on human rights and other legal obligations of States. For comparison, in 2018, the ASEAN States called for a climate change-resilient region in the aftermath of Tropical Storm Tembin that affected the Philippines. In a statement, the ASEAN Secretary-General highlighted that the area has to enhance the resilience of our communities through disaster risk reduction interventions, climate change adaptation strategies, as well as conservation and sustainable management of biodiversity and natural resources (ASEAN, 2018). There is a need for legal and policy frameworks that enable climate change mitigation and adaptation. These can include environmental rights laws and facilitating investment in climate change technologies.

Regional bodies need to be proactive in preventing further climate damage and law reforms can be one of the methods of arresting such harm by providing for tangible rights for vulnerable population groups (Obydenkova, 2022). Functional human rights and climate change monitoring systems in vulnerable regions can be instrumental in safeguarding legal rights in times of climate change. The systems could be used to strengthen peer review mechanisms among States and to give ordinary people the space to come up with climate change resilience mechanisms. The legal implications of climate change and the attendant harms have implications beyond the borders of a single nation-state (Carter et al., 2021).

3.2 *National Legal Responses to Climate Hazards in Zimbabwe*

Attention to developing States such as Zimbabwe is essential and constitutes a form of climate justice since these States emit minimal volumes of greenhouse gases (GHG) in contrast with developed industrialised States. Zimbabwe, for instance, contributes "a mere 1.7 per cent to the total GHG emissions on the African continent" (Zimbabwe's National Climate Change Response Strategy, 2014, p. 42). However, it is also critical to point out that in line with the Paris Agreement's "common but differentiated responsibilities" (Paris Agreement, art 2) policy, Zimbabwe's National Climate Change Response Strategy indicates that "it is still important for future development planning to take cognisance of this low carbon footprint and to preserve it" (Zimbabwe's National Climate Change Response Strategy, 2014, p. 42). Hence, sub-Saharan African states are taking responsibility even though their contribution to GHG emissions is almost negligible.

Zimbabwe has been somewhat proactive in coming up with climate, environmental and disaster response laws and policies. As of May 2023 however, the nation was yet to compile and enact its various climate policies and funds into a comprehensive Climate Change Act. In terms of alignment with global environmental and climate governance, the state is a signatory to some of the key multilateral environmental agreements available. Zimbabwe is a signatory to, among others: the United Nations Framework Convention on Climate Change, the Kyoto Protocol, the Montreal Protocol on Substances that Deplete the Ozone Layer, the United Nations Convention on Biological Diversity (UNCBD), the Convention on International Trade in Endangered Species of Wild Fauna and Flora (CITES) and the United Nations Convention on the Law of the Sea. Zimbabwe is also one of the States instrumental in the development of the SADC Climate Change Adaptation (CCA) Strategy for the Water Sector.

Zimbabwe has also played a proactive role in formulating climate change policy at the domestic level (Zimbabwe Revised Nationally Determined Contribution, 2021). The State is a signatory to the Paris Agreement and essential treaties on the environment and climate change. The State also formulated Zimbabwe's National Climate Change Response

Strategy, the Zimbabwe Climate Policy and the National Energy Policy. The State framed the Response Strategy intending to guide Zimbabwe's "response measures in addressing the impacts of climate change" (Zimbabwe's National Climate Change Response Strategy, Foreword). The Response Strategy highlights that the authorities in Zimbabwe are aware of the various adverse impacts of climate change. The Response Strategy states:

> Research in Zimbabwe has revealed that the majority of rural Zimbabweans live in semi-arid zones, and will suffer disproportionately from the emerging impacts of climate change and variability including disasters associated with extreme weather events such as droughts, periodic flooding, disease outbreaks for both human and livestock and loss of croplands. (*Zimbabwe's National Climate Change Response Strategy*, Foreword by Hon Saviour Kasukuwere, Minister of Environment, Water and Climate.)

Despite strong-worded commitments, a viable and adequate response to climate-induced disasters is lacking on the ground in Zimbabwe (Munsaka et al., 2021). This is a major challenge for vulnerable population groups who suffer from the climate crisis and desire practical solutions instead of words without effective action in terms of budgetary support and specific law enactments. Global South states like Zimbabwe and Mozambique must overcome the mismatch to protect the human and environmental rights of their respective citizenry in times of climate change.

The *Zimbabwe Climate Policy* is another example of a state policy that is aware of the impacts of climate change mitigation measures, especially on agriculture, which is one of the country's three main economic sectors besides mining and the service industry. Agriculture is also important since a large section of the vulnerable population groups in Zimbabwe rely on rain-fed subsistence farming for food security. In the Climate Policy, for instance, the State commits to "ensure that mitigation and adaptation measures enhance agriculture-based livelihoods, by promoting food security and poverty alleviation" (Zimbabwe Climate Policy, p. 11).

The main pieces of legalisation that deal with climatic hazards in Zimbabwe, in the absence of a viable climate enactment as at the time of this writing, are the Constitution of Zimbabwe (2013), the Environmental Management Act and the Civil Protection Act. The latter particularly was enacted for the purposes of establishing a civil protection organisation in Zimbabwe as well as to provide for the operation of civil protection services in times of disaster. Furthermore, the Act provides for the establishment of a fund to finance civil protection. The legislation however outdated is in operation in Zimbabwe and Mozambique to deal with disasters, though these may have been enacted before the current frequency of climatic hazards.

The Civil Protection Act in Zimbabwe requires an amendment to better equip and fund civil protection departments and to ensure that climatic hazards do not always transform into disasters. This and the other relevant statutes highlight that Zimbabwe has reasonably comprehensive environmental legislation but lacks adequate implementation in terms of budgetary support and complement institutions. This leads to great vulnerability among the affected sections of the population (Chanza et al., 2020). Hence, the world witnessed over 600 lives being lost to Cyclone Idai alone as the state failed to timeously respond or to introduce early warning systems to families located in the line of fire.

Various laws are relevant in the climate discourse in Zimbabwe including the Environmental Management Act and the Civil Protection Act. The Civil Protection Act in particular needs to be reformed and funded to deal with the increasingly adverse impacts of the climate crisis. More importantly, Zimbabwe needs to come u with a comprehensive Climate Change Act that provides for such aspects such as a statutory Climate Fund.

3.3 *National Legal Responses to Climate Hazards in Mozambique*

Mozambique has been at the receiving end of various serious challenges in its history, including political crises, civil wars, terrorism, inequality and environmental hazards. In terms of environmental disasters, the country's geographical location and economic challenges contribute to the extent of the hazards. Mozambique is a coastal state with 2500 kilometres of coastline along the Indian Ocean which makes it a target for tropical cyclones coming from Madagascar and the Mozambique Channel. Being a country that will suffer from climatic hazards, there should be greater emphasis and resources targeted at mitigation and adaptation. As Foley points out, "it is beyond the power of Mozambique's government to influence the level of climate change, so the main question must concern its ability to cope with the consequent effects" (Foley, 2007, p. 6).

In terms of legal and public policy responses, in the 1990s and early 2000s, the rest of the world was becoming conscious of the impact of the environmental crisis. Mozambique enacted some environmental-related policies such as the National Climate Change Strategy, the 2006 National Master Plan for Prevention and Mitigation of Natural Calamities and legislation such as the Environmental Law Act No. 20 of 1997. The National Climate Strategy emphasises the need to improve climate disaster preparedness through measures that include relocation and protection of vulnerable people and property. The strategy also discusses the need to strengthen the role of the disaster response council in coordinating operations, evacuation, relief, reconstruction and support for climate disaster victims in Article 4.6.1.1.2.

The Environmental Law Act 20/97 outlines, in sections 19–24, the environmental rights and duties of citizens. The Act provides for access to information in section 19 concerning the environment as well as the right to education in section 20, to better inform and encourage public participation in environmental decision-making. These are important rights to ensure communities are made aware of climatic hazards before they are impacted. However, Mozambique, like Zimbabwe, has not enacted specific climate legislation at the moment. Practical responses from the Mozambique authorities have mostly been in the form of policies, decrees and reports.

The State is committed to addressing the impact and causes of climate change. On climate mitigation, Mozambique enacted Decree No. 24 of 2008 which approves the Regulation on the management of substances damaging the ozone layer. In terms of climate disaster response, the State has yet to enact a specific law. However, there are laid out institutions tasked to manage disaster response, operating similar to Zimbabwe's Civil Protection Department. In Mozambique, such institution is the *Conselho Nacional Técnico de Gestão de Calamidades Naturais*[1] (CTGCN). Furthermore, reports showing intention to address the climate crises have been drafted in line with Mozambique's international commitments to United Nations bodies. The Fifth National Report to the Convention on Biological Diversity clearly outlines the policy desire by Mozambique to acknowledge and highlight the intention of the Mozambique government to deal with the impacts of climate change:

[1] English: National Disaster Management Technical Council.

> Climate change is of concern; it may lead to an increased vulnerability of certain ecosystems (for example by enhancing the magnitude and the frequency of extreme events such as droughts, floods and cyclones or other events like wildfires or coastal erosion) leading to the loss of biodiversity and some habitat areas if effective measures are not undertaken. (Ministry for The Coordination of Environmental Affairs, Fifth National Report on the Implementation of the Convention on Biological Diversity in Mozambique, 2014, p. 15)

This report is critical in the Mozambique context. Cyclones have resulted in the loss of lives, properties, infrastructure and livelihoods on several occasions in twenty-first Century Mozambique. In terms of response, the country adopted a National Climate Change Strategy which runs from 2013 to 2025 in addition to other measures from state and non-state actors.

Civil society plays a key role in shaping the governance, and legal and policy responses in Mozambique. The role of environmental civil society in Mozambique is mirrored beyond the country. One of the critical climate litigation matters is the case of *Friends of the Earth v. UK Export Finance CO/3206/2020*. The case was brought in late 2021 before courts in the United Kingdom and concerned the climate impacts of investments in Mozambique's natural gas fields. The case was brought in the UK following an application in September 2020 by the environmental group, Friends of the Earth England Wales and Northern Ireland. The group made the legal challenge against the decision by the UK Export Finance to provide over US$1 billion of public funds money to finance a TotalEnergies liquefied natural gas in Mozambique. The project is for the extraction of 43 million tonnes of LNG per year for 32 years. The argument was that this would result in total combustion emissions of 4.3 billion tonnes of carbon emissions.

Mozambique's history has been interspersed with various hazards: man-made, environmental and conflict-related. The conflicts thus shape the history of the country, especially in contemporary times when conflict is exasperating for communities already in the throes of the climate crisis. Conflicts in Mozambique have ranged from ideological ones to the recent wars in natural gas-rich provinces such as Tete and Cabo Delgado Provinces. These conflicts arguably have underlying social inequality causation (Darch, 2016). In terms of causing conflict, climate and environmental issues have also been to blame. Climate injustice in itself is practically one of the foremost culprits behind conflicts around the world. For illustration, across the oceans in South East Asia, notable climate-related disasters have affected South East Asia such as the Indian Ocean Tsunami (2008), Cyclone Nargis (2008), and Typhoon Haiyan (2013). Cyclone Nargis reportedly killed 100,000 people in Myanmar alone (National Geographic, 2008). Myanmar is a country that has encountered years of natural resource conflict between the government and ethnic minorities such as the Rohingya. Simpson points out that conflict usually has a toll on environmental justice:

> Environmental governance and protection is often subordinated to economic growth, particularly under illiberal regimes, and this pattern has played out across Asia in recent decades. Myanmar has been no exception to this trend. The unequal division of access to resources and capital that accompanied half a century of authoritarianism has been reflected in the profitable exploitation of the environment by the few, with the costs of the resultant environmental degradation borne by the many. (Simpson, 2017, p. 161)

Hence, environmental injustice and socio-economic conflict can be seen as significant contributors to the human rights crises affecting vulnerable populations around the world, especially in the Global South in which both Mozambique and Zimbabwe are a part, as is Myanmar.

It is highly relevant for the Mozambique climate strategy mentioned above to discuss ways of improving the response of local vulnerable communities to climate change. These are some of the basic tenets of climate justice. Furthermore, Mozambique is open to climate-related impacts due to its geographical location, hence it should continuously ratify and develop laws and policies that provide resilience to the shocks of climate change for its vulnerable population groups.

4 Comparative Legal and Policy Analyses

The challenges brought about by climate change are not unique to Southern Africa. Also common are the legal and social justice implications of environmental changes. A look at global environmental events has made it obvious that there is an indelible link between climate change, its impacts and social and class justice (Levy & Patz, 2015). These events, therefore, call for Southern African states to adopt a comparable and collaborative approach in the development of their responsive climate legal and policy frameworks.

Both Zimbabwe and Mozambique are part of the SADC Climate Services Centre. Both States have encountered some of the worse climate and non-environmental impacts in recent years. The legal responses in the region have however varied in historical terms. A case in point is the Republic of South Africa which outlined in its 2004 Policy that:

> It is current thinking that a specific Act of Parliament to deal with climate change issues in South Africa is not warranted by present circumstances. However, specific amendments can be made to other laws that embrace environmental issues such that climate change provisions are included. (Department of Environmental Affairs and Tourism, 2004, para 4.8)

Fast forward to 2022, and a Climate Change Bill for the Republic of South Africa is in its advanced stages at the time of writing this chapter. The Bill seeks to enable the development of an effective climate change response and a long-term, just transition to a low-carbon and climate-resilient economy and society for South Africa in the context of sustainable development.

Regional issues aside, both Zimbabwe and Mozambique have ratified some international human rights treaties. The Constitution of Zimbabwe (2013) imposes a duty on the State to "ensure that all international conventions, treaties and agreements to which Zimbabwe is a party are incorporated into domestic law" (Constitution of Zimbabwe, sec 34). Among the treaties that are relevant to the protection of human rights in whatever circumstances, it is pertinent to mention that Zimbabwe has ratified the Convention on Economic, Social and Cultural Rights (CESCR), the African Charter on Human and Peoples Rights and its protocols, the Convention on the Elimination of All Forms of Discrimination Against Women (CEDAW) and the Convention on the Rights of the Child (CRC).

In light of the above, Zimbabwe and Mozambique have a special legal, social and moral obligation towards rural communities affected by poorly regulated and poorly-planned climate change response mechanisms.

At the national level, Zimbabwe has relatively more comprehensive lawsthan Mozambique. In terms of municipal laws, Zimbabwe has a more comprehensive climate and disaster legal framework. The major challenge is in the implementation and funding of the existing laws. These include the Constitution of Zimbabwe, especially section 73, the Environmental Management Act, and the Civil Protection Act. In terms of policy, the key policies include the Climate Change Policy and the National Climate Change Response Strategy.

The 2013 Constitution of Zimbabwe has a more coherent Declaration of Rights in contrast with the previous 1980 independence Constitution.

Like in many other provisions dealing with social and economic rights, the Constitution includes the statement—"within the limits of the resources available"—to most of these rights (Constitution of Zimbabwe, Declaration of Rights). There are, however, no clear steps established to measure and monitor the resources regarded as being available to the State. Furthermore, there are no demonstrable commitments by the State to report on progress made towards the implementation of such rights.

5 Conclusion

Climatic hazards are set to increase with the ongoing environmental, social and economic crisis. Hence viable, innovative and practical legal and policy frameworks are a necessity for developing countries in the Global South in general and sub-Saharan Africa in particular. One key strategy in climate response policy-making is enacting or updating relevant laws relating to disaster responses, climate justice and environmental rights. There is a need to improve research that can provide data and knowledge for effective and innovative climate response policy and law-making. This has been partly attained in Mozambique with the establishment of a Centre of Knowledge on Climate Change (CGC) within the Ministry of Science and Technology and with the Climate Change Management Department within the Ministry of Environment, Climate, Tourism and Hospitality Industry in Zimbabwe. The laws and policies of both states relating to the disastrous impacts of climate and environmental hazards need extensive review and updating especially on emergency response, public funding, human rights protections and environmental justice. Day-to-day laws and policies worldwide need to be adaptive to the climate crisis and climate change. In addition to municipal actions, Global South states like Zimbabwe and Mozambique need climate financing, legal, policy and technological support from the Global North in line with the Sendai Framework, Kyoto protocol, UNFCCC and other multilateral environmental agreements.

References

Abouyoub, Y. (2012). Climate: The forgotten culprit. The ecological dimension of the darfur conflict. *Race, Gender & Class, 19*(1/2), 150.

Adelman, S., & Lewis, B. (2018). Rights-based approaches to climate change. *Transnational Environmental Law, 7*(1), 9.

ASEAN, Statement of the Secretary-General of ASEAN, H.E. Dato' Lim Jock Hoi, on the Landfall of Tropical Storm Tembin (Vinta) (Jakarta: ASEAN, 03 January 2018).

Atapattu, S. (2009–2010) Climate change, human rights, and forced migration: Implications for international law. *Wisconsin International Law Journal, 27*(3), 607–636.

Bellinkx, V., Casalin, D., Erdem Türkelli, G., Scholtz, W., & Vandenhole, W. (2022). Addressing climate change through international human rights law: From (extra)territoriality to common concern of humankind. *Transnational Environmental Law, 11*(1), 69–93.

Berbés-Blázquez, M., Mitchell, C. L., Burch, S. L., & Wandel, J. (2017). Understanding climate change and resilience: Assessing strengths and opportunities for adaptation in the Global South. *Climatic Change, 141*, 227–241.

Bodansky, D., Brunnée, J., & Rajamani, L. (2017). *International climate change law*. Oxford University Press.

Bopape, M. J. M., et al. (2019). A regional project in support of the SADC cyber-infrastructure framework implementation: Weather and climate. *Data Science Journal, 18*(34), 1–10.

Carter, T. R., Benzie, M., Campiglio, E., Carlsen, H., Fronzek, S., Hildén, M., Reyer, C. P. O., & West, C. (2021). A conceptual framework for cross-border impacts XE "impact" of climate change. *Global Environmental Change, 69*, 1–14.

Chanza, N., Siyongwana, P. Q., Williams-Bruinders, L., et al. (2020). Closing the gaps in disaster management and response: Drawing on local experiences with cyclone Idai in Chimanimani, Zimbabwe. *International Journal of Disaster Risk Science, 11*, 655–666.

Charter of the United Nations. (Adopted 26 June 1945, entered into force 24 October 1945) 1 UNTS XVI (UN Charter).

Chowdhury, A. R., & Bhuiyan, J. H. (2010). *An introduction to international human rights law*. BRILL.

Civil Protection Act [Chapter 10:06] (Zimbabwe).

Climate Change Act, 2008 (United Kingdom).

Climate Change Act, 2016 (Kenya).

Constitution of Zimbabwe [Amendment (No 20) Act, 2013].

Convention on the Elimination of All Forms of Discrimination Against Women (adopted 18 December 1979, entered into force 3 September 1981) 1249 UNTS 1 (CEDAW).

Convention on the Rights of the Child (adopted 20 November 1989, entered into force 2 September 1990) 1577 UNTS 3 (CRC).

Cubie, D. (2019). *The international legal protection of persons in humanitarian crises: Exploring the acquis humanitaire*. Bloomsbury Academic.

Darch, C. (2016). Separatist tensions and violence in the 'model post-conflict state': Mozambique since the 1990s. *Review of African Political Economy, 43*(148), 320–327.

Department of Environmental Affairs and Tourism. (2004, September). *A national climate change response strategy for South Africa*. Department of Environmental Affairs and Tourism.

Donnelly, J. (2007). The relative universality of human rights'. *Human Rights Quarterly, 29*, 281.

Draft Africa Climate Change Strategy 2020–2030.

Environment Law Approved 28/7/97 (Mozambique).

Fershee, J. P. (2009). The rising tide of climate change: What America's flood cities can teach us about energy policy, and why we should be worried. *Environmental Law, 39*(4), 1109–1141.

Fifth National Report on the Implementation of the Convention on Biological Diversity in Mozambique (Ministry for The Coordination of Environmental Affairs Maputo 2014).

Foley, C. (2007). *Mozambique: A case study in the role of the affected state in humanitarian action* (HPG Working Paper).

Friends of the Earth v. UK Export Finance CO/3206/2020.

Godden, N. J., Wijekoon, D., Wrigley, K., et al. (2022). Social (In)justice, climate change and climate policy in Western Australia. *Environmental Sociology, 8*(4), 377–387.

Government of Mozambique, *National Climate Change Adaptation and Mitigation Strategy* (Approved during the 39th Session of the Council of Ministers: Maputo, Mozambique 13th November 2012).

Government of Zimbabwe. (2012). *National Energy Policy*. Government of Zimbabwe.

Government of Zimbabwe. (2014). *Zimbabwe's National Climate Change Response Strategy*. Government of Zimbabwe.

Government of Zimbabwe. (2021). *Zimbabwe Revised Nationally Determined Contribution*. Government of Zimbabwe.

Hayes, K., Blashki, G., Wiseman, J., Burke, S., & Reifels, L. (2018). Climate change and mental health: Risks, impacts XE "impact" and priority actions. *International Journal of Mental Health Systems, 12*, 28. https://doi.org/10.1186/s13033-018-0210-6

IPCC. (2022). Climate Change 2022: Impacts, adaptation, and vulnerability. In H.-O. Pörtner, D. C. Roberts, M. Tignor, E. S. Poloczanska, K. Mintenbeck, A. Alegría, M. Craig, S. Langsdorf, S. Löschke, V. Möller, A. Okem, & B.

Rama (Eds.), *Contribution of Working Group II to the Sixth Assessment Report of the Intergovernmental Panel on Climate Change*. Cambridge University Press.

Jordan, A., & Huitema, D. (2014a). Innovations in climate policy: Conclusions and new directions. *Environmental Politics, 23*(5), 906–925.

Jordan, A., & Huitema, D. (2014b). Policy innovation in a changing climate: Sources, patterns and effects. *Global Environmental Change, 29*, 387–394.

Koch, M. (2011). *Capitalism and climate change: Theoretical discussion, historical development and policy responses*. Palgrave Macmillan.

Levy, B. S., & Patz, J. A. (2015). Climate change, human rights, and social justice. *Annals of Global Health, 81*(3), 310–322.

Maluwa, T. (2012). Ratification of African Union treaties by member states: Law, policy and practice. *Melbourne Journal of International Law, 13*(2), 636–684.

Ministry of Environment, Climate, Tourism and the Hospitality Industry. (2014). *Zimbabwe's National Climate Change Response Strategy.*

Munsaka, E., Mudavanhu, C., Sakala, L., et al. (2021). When disaster risk management systems fail: The case of Cyclone Idai in Chimanimani District, Zimbabwe. *Int J Disaster Risk Sci, 12*, 689–699.

National Geographic, '100,000 May Have Died in Myanmar Cyclone, U.S. Says' National Geographic News (Washington D.C.: 07 May 2008) Retrieved from https://news.nationalgeographic.com/news/2008/05/080507-AP-myanmar-cyc.html. Accessed 26 December 2017.

Nhamo, G., & Chikodzi, D. (2021). *Cyclones in Southern Africa: Volume 3: Implications for the sustainable development goals*. Springer Nature.

Nhamo, G., & Chikodzi, D. (2021) *Cyclones in Southern Africa Volume 1: Interfacing the catastrophic impact of Cyclone Idai with SDGs in Zimbabwe*. Springer.

Nyahunda, L., Chibvura, S., & Tirivangasi, H. M. (2021). Social work practice: Accounting for double injustices experienced by women under the confluence of covid-19 pandemic and climate change impacts in Nyanga, Zimbabwe. *Journal of Human Rights and Social Work, 6*(3), 213–224.

Obydenkova, A. (2022). Sustainable development and actors of regional environmental governance: Eurasia at the crossroads. *Problems of Post-Communism, 69*(4–5), 436–443.

Panke, D., & Stapel, S. (2022). Architects of regional regime complexity: States and regional organizations in Europe. *Journal of Contemporary European Studies, 30*(3), 387–404.

Paris Agreement (adopted 12 December 2015, entered into force 4 November 2016).

Pegasys, A. (2014). *Climate change adaptation: perspectives for the Southern African Development Community SADC: Long term adaptation scenarios*. Pegasys.

Percival, R. V., & Zhao, H. (2014). The role of civil society in environmental governance in the United States and China. *Duke Environmental Law & Policy Forum, 24*, 142. (U of Maryland Legal Studies Research Paper No. 2014-36).

Schulz, M. (2001). *Regionalization in a globalizing world*. Zed Books.

Sendai Framework for Disaster Risk Reduction 2015–2030 (Adopted 3 June 2015) A/RES/69/283.

Setzer, J., & Benjamin, L. (2020). Climate change litigation in the global South: Filling in gaps. *AJIL Unbound, 114*, 56–60.

Simpson, A. (2017). Starting from year zero: Environmental governance in Myanmar. In S. Mukherjee & D. Chakraborty (Eds.), *Environmental challenges and governance: Diverse perspectives from Asia* (pp. 152–165). Routledge.

Söderbaum, F. (2013). Rethinking regions and regionalism. *Georgetown Journal of International Affairs, 14*(2), 9–18

Statement of the Secretary-General of ASEAN, H.E. Dato' Lim Jock Hoi, on the Landfall of Tropical Storm Tembin (Vinta) (Jakarta: ASEAN, 03 January 2018).

The African Union, 'Draft African Union Strategy on Climate Change' AMCEN-15-REF-11.

United Nations Declaration on the Rights of Peasants and Other People Working in Rural Areas (Adopted on 28 September 2018) A/HRC/RES/39/12, (UNDRIP).

Universal Declaration of Human Rights (adopted 10 December 1948) UNGA Res 217 A (III) (UDHR).

World Bank. (2019, April 11). *Statement on high-level meeting on humanitarian and recovery efforts following Cyclone Idai*. World Bank.

Zungu, L. T., Greyling, L., & Mbatha, N. (2021). Economic growth and income inequality: A non-linear econometrics analysis of the SADC region, 1990–2015. *African Journal of Economic and Management Studies, 12*(2), 285–301.

CHAPTER 8

Governing Urban Green Spaces in the Context of Climate Change in Ghana

Nelson Nyabanyi N-yanbini

1 Introduction

Climate change, perhaps, is the most protracted challenge of the twenty-first century, threatening livelihoods and exacerbating vulnerability of cities. Evidence of the impact of climate change on cities abounds (Asibey & Cobbinah, 2022). Extreme weather events in cities such as urban heat waves, flooding, drought, and sea level rise are widespread manifestations of climate change (Cobbinah et al., 2019). In most climate change disasters, cities turn to suffer the greatest impacts due to rising vulnerabilities amidst low adaptation (IPCC, 2019). Climate change affects people and cities differently due to differences in adaptive capacities. As a result, the poor and vulnerable populations in the global south suffer the most impacts of climate change, despite contributing the least. Adaptation to climate change is thus, essential, with recent research

N. N. N-yanbini (✉)
Department of Urban Design and Infrastructure Studies, SD Dombo University of Business and Integrated Development Studies, Wa, Ghana
e-mail: nelsonnyanbini@gmail.com

M. Addaney et al. (eds.), *Climate Change in Africa*,
https://doi.org/10.1007/978-3-031-30050-9_8

showing strong evidence of the need for urban green spaces (UGS) in climate change adaptation.

Globally, the contribution of UGS to climate change adaptation and urban resilience is enormous (Kitha & Lyth, 2011; van der Jagt & Lawrence, 2019). Urban green spaces are defined as public or private outdoor urban amenity that provide environmental, social, and economic benefits (Wendel et al., 2012). This could be natural, quasi-natural, and/or human-modified landscapes (Etingoff, 2016; Pearlmutter et al., 2017). UGS include parks, gardens, recreation venues, vacant plots, brownfields, and general greenery, often sited within the city, along water bodies such as lakes, rivers, along transportation routes like railway corridors, roads, and around historical sites (Gupta et al., 2012).

UGS offer unique ecosystem services with distinctive functions within urban areas. For instance, it makes urban areas livable by conserving biodiversity, enhances mental and physiological well-being, and balances urban heat islands (Kendal et al., 2016; Mayrand et al., 2018). Beyond this, UGS contribute to improvement in air quality, reduce outdoor temperatures by creating shade, increase carbon sink, and reduce flooding (Govindarajulu, 2014). This makes UGS ideal "soft engineering" strategy in climate change adaptation (Kitha & Lyth, 2011, p. 251).

The governance of UGS, however, is complex and comes with diverse styles which incorporate different spaces, land use/tenure types, and cultural underpinnings (Amoako & Adom-Asamoah, 2019). For this reason, Mensah et al. (2015) argued for collaborative and all-inclusive governance of UGS. Good governance of UGS involve working with and for diverse groups with different interests on land uses (Coffey et al., 2020; Lockwood et al., 2010). Good governance principles developed by Lockwood et al. (2010) were applied to the selected cases to understand the governance of urban greening initiatives.

Even though many factors shape the provision of UGS, its preservation has political, socio-cultural, and economic implications (Safransky, 2014). Due to its intricate and seemingly fragile context, sustainable governance of UGS is essential toward the resilience of cities (Coffey et al., 2020; Lindley et al., 2018). However, not much is known about the governance of UGS as adaptive responses to climate change, especially in the global south.

This chapter uses a case study design in theorizing and understanding the governance of UGS as adaptive responses to climate change. The initial scoping involves a review of climate change, governance, and UGS concepts with emphasis on good governance and sustainability principles (Lockwood et al., 2010). This was followed by context analysis of four urban greening initiatives in Ghana, implemented by both public and private agencies in efforts to revitalize green spaces. The selected cases transcend different tenure arrangements, scales, and governance with a complex mix of stakeholders and interests (Coffey et al., 2020; Lawrence et al., 2013; Lemos & Agrawal, 2006; Lockwood, 2010). The four selected cases were: Green Ghana Project (GGP), The Green Republic Project (TGRP), Achimota Forest (AF), and the Accra Urban Green Infrastructure Project (AUGIP).

2 Literature Review

2.1 *Urbanization, Climate Change, and the Role of Urban Green Spaces*

Urban areas have become the focal point of many planning interventions. This is because the world's urban population is projected to increase from 55% in 2018 to 68% by 2050 (United Nations, 2018). Ghana's urban population has increased nearly 12-fold from 1.5 million in 1960 to 17.6 million in 2020. Unsustainable urbanization has led to rising levels of energy consumption, increase imperviousness, heat waves, carbon emissions, and general environmental deterioration, threatening the sustainability of cities (Mayrand et al., 2018; Mensah et al., 2015; Reckien et al., 2014; Sturiale, 2019). In the midst of this urban growing trend, global leaders formulated the Sustainable Development Goals (SDGs) in 2015 to build resilience of cities to adapt to climate change impacts. This is particularly important in achieving Goal 11, sustainable cities and communities and Goal 13, climate action, in building resilient, safe, and inclusive cities (United Nations, 2020). The impact of climate change on cities has been well documented, as urban disaster risks are now common in cities.

The rising impacts of climate change has threatened the sustainability of urban ecosystems, exposing cities to all forms of vulnerabilities, including rising levels of urban poverty, flood risks, heat waves, droughts, and associated disasters. In Ghana, temperatures are expected to rise by 2.0 °C by 2050 and 3.9 °C by 2080 (Cobbinah & N-yanbini, 2019). This will further exacerbate flooding, urban heat, drought, sea level rise, and erratic rainfall patterns, leading to declining agriculture production and unsustainable energy consumption.

Measures to contain climate change have come in twofold: mitigation and adaptation. Mitigation aims to reduce carbon emissions and increase carbon sink while adaptation seeks to build resilience by modifying people's lives and cities to sustainably respond to the impacts of climate change. By resilience, I mean the adaptive capacities of cities to absorb the impacts of climate change while maintaining its functionality. Despite the importance of hard engineering solutions to climate change, soft engineering measures which come in the form of adaptation, such as green spaces, open spaces, and urban landscapes have a myriad of ecological and social benefits. Therefore, using UGS as adaptation to climate change comes at minimal cost and nature-based, making it ideal for developing countries (Kitha & Lyth, 2011).

Approaching climate change through UGS has many benefits in improving the adaptive capacity of cities. First, it reduces urban heat islands effect through a reduction in urban temperatures. Also, UGS enhance hydrology, leading to reduced surface runoff (Gill et al., 2007). UGS act as buffers, during extreme flood events, hence serving as natural stormwater drains (Govindarajulu, 2014). This helps to reduce climate-weather related risks as a result of extreme events. UGS reinforce the process of carbon sequestration, thereby serving as a climate change mitigating measure (Escobedo et al., 2011; Govindarajulu, 2014). UGS increase carbon sink, as a form of building resilience of urban areas to respond to climate change (Sturiale, 2019). The need for urban greening policies to give direction and serve as the benchmark through which progress can be measured is desired.

While some urban areas have braced up the challenge and institutionalized measures to respond to the impacts of climate change through UGS, others are taking spontaneous and unsustainable measures. For example, Kampala city authority in Uganda has developed a policy document; Urban Greening Infrastructure Ordinance (UGIO of 2019) as a response to climate change through urban greening. The UGIO policy requires all public spaces to have a tree for every 120 square meters. For private plots, the policy requires 40% of green space and a tree for every 80 square meters. Commercial and industrial areas were tasked to have a minimum of 10% to 35% of green spaces for every development in Kampala city. This policy makes comprehensive efforts toward urban greening restoration in Kampala city.

Another example is Durban, the first city in South Africa to integrate open space system into its urban planning schemes (Shih, 2017). Durban city is governed by urban greening policies designed by urban and regional planners in collaboration with city authorities. Durban is noted for its excellent combination of green spaces and amenity, creating an aesthetically pleasing environment (Robberts, 1994; Shih, 2017). Beyond ecological relevance, Durban city authorities have expanded the open space system to include climate co-benefits such as community-based ecosystems, and the provision of jobs for local livelihoods (eThekwini Municipality and ICLEI Africa Secretariat, 2007). These innovations have helped to sustain Durban's green spaces. Although other cities in Africa such as Johannesburg (South Africa), Kigali (Rwanda), and Accra (Ghana) are in the transition toward green economies, such transitions should be backed by green space policy toward urban sustainability.

2.2 *Conceptualizing the Governance of Urban Green Spaces*

Governance is akin to the structures and processes through which decisions are made and power shared in running the affairs of a community, society, or country (Schultz et al., 2015). Governance is thus linked to the powers of the state, civil society groups, and private businesses (Lemos & Agrawal, 2006), making decisions at various levels. Governance has two major components: the structures and the processes through which these structures work. The selected urban greening initiatives cut across different spatial scales, involving stakeholders in business, government, community actors, and civil society organizations (Lawrence et al., 2013; Lemos & Agrawal, 2006).

Urban greening decisions affect individuals at the household levels through to community and the larger society. As a result, governance of urban greening must identify the various interest groups and make adequate provisions for same in order to achieve good results. The relationship between structures, the different interests and how it affects the governance of urban greening are thus essential (Diko & Hollstein, 2021). Based on the elements of actors, power, and interactions, Arnouts et al. (2012) described four types of governance arrangements. These include hierarchical governance, closed co-governance, open co-governance, and self-governance (Coffey et al., 2020).

However, good governance of urban green spaces as an ecosystem-based approach to climate change adaptation has received limited research attention in the global south. The need for sustainable environmental governance is essential in the wake of increasing impacts of climate change (Lockwood et al., 2010). The management of natural resources fall within environmental governance cluster, requiring interactions with complex mix of stakeholders including public, private, and civil society groups. The governance of urban greening, therefore, coordinates this complex mix of interests, stakeholders, and organizations at different spatial and temporal scales. This requires pragmatic approaches to liaise with the many interest groups while maintaining the environmental focus of the initiatives in the governance framework. Good governance principles according to Lockwood et al. (2010) thus help to examine the contributions of urban greening initiatives to environmental governance (Table 1).

The complexity in managing green spaces often come with overlapping and sometimes conflicting interests among stakeholders (Amoako & Adom-Asamoah, 2019). These conflicts arise as a result of the variations in the coordination and management of green spaces. Due to the different power hierarchies in the governance of settlements in Africa, it is difficult to usurp some of these interests by urban planning decisions (Amoako & Adom-Asamoah, 2019). Lack of cooperation between urban governance stakeholders and urban planners has dire consequences for loss of urban green spaces in Africa (Mensah et al., 2015). For instance, residential properties have been developed on urban green spaces or ecologically sensitive areas in many towns in Africa (Mensah et al., 2015). This trend of development has claimed many acres of green spaces, making urban areas unsustainable. This has fueled recent concerns about the governance of urban green spaces toward building resilience.

3 Research Methodology

This research adopts a case study design (Yin, 2014) to analyze the governance of urban greening initiatives as a process in environmental governance through the application of good governance principles. The criterion for inclusion was that, each selected initiative should have implemented at least one greening project in an urban area, or the entire initiative should have urban greening focus. This led to the selection of four (4) initiatives with two having entirely urban green focus while the other two have at least one urban greening project among its goals. The four (4) urban greening initiatives were selected as cases based on which analysis was done. These initiatives include the Green Ghana Project

Table 1 Good governance principles

Principle	*Explanation*
Legitimacy	'(a) the validity of an organization's authority to govern that may be (i) conferred by democratic statute; or (ii) earned through the acceptance by stakeholders of an organization's authority to govern; (b) that power being devolved to the lowest level at which it can be effectively exercised; and (c) the integrity with which this authority is exercised' (p. 991)
Transparency	'(a) visibility of decision-making process; (b) the clarity with which the reasoning behind decision is communicated; and; (c) the ready availability of relevant information about governance and performance in an organization' (p. 993)
Accountability	'(a) the allocation and acceptance of responsibility for decision and actions and (b) the demonstration of whether and how these responsibilities have been met' (p. 993)
Fairness	'(a) the respect and attention given to stakeholders' views; (b) consistency and absence of personal bias in decision-making; and (c) the consideration given to the distribution of costs and benefits of decisions' (p. 994)
Inclusiveness	'Opportunities available to stakeholders to participate in and influence decision-making processes and actions' (p. 993)
Integration	'(a) the connection between, and coordination across, different governance levels; (b) the connection between, and coordination across, organizations at the same level of governance; and (c) the alignment of priorities, plans, and activities across governance organizations' (p. 995)
Capability	'The systems, plans, resources, skills, leadership, knowledge, and experiences that enable organizations, and the individuals who direct, management, and work for them, to effectively deliver on their responsibilities' (p. 996)
Adaptability	'(a) the incorporation of new knowledge and learning into decision-making and implementation; (b) anticipation and management of threats, opportunities, and associated risks; and (c) systematic reflection on individual, organizational, and system performance' (p. 996)

Source Adopted from Lockwood et al. (2010) and Coffey et al. (2020)

(GGP), Achimota Forest (AF), The Green Republic Project (TGRP), and the Accra Urban Green Infrastructure Project (AUGIP). The selected cases cut across people with diverse social, economic, cultural, and political backgrounds. Again, the selected cases transcend different actors; civil society groups, policymakers, private individuals, and public workers, all of who put efforts at different levels to increase Ghana's tree canopy.

The projects cut across different scales, with varying levels of stakeholder involvement, administration, governance styles, and structures.

As part of the initiatives to restore environmental cover, the Government of Ghana declared Green Ghana Day; an initiative under the Forestry Commission as part of the Green Ghana Project (GGP), which led groups to plant trees at strategic points such as institutions, backyards, school compounds, residences, and areas 100 meters away from road reservations. The Green Ghana Project resulted in the planting of some seven million trees in its first year. In the second year of implementation, GGP planted 20 million trees with efforts to revitalize Ghana's green spaces. This project is one of the innovative ways of sustaining green infrastructure since it adds to Ghana's tree canopy at the macro-level. Achimota Forest (AF) is one of the oldest and largest urban green spaces in Accra, the capital city of Ghana. AF, covering 494.95 hectares of land, was gazetted in 1930 by Sir Frederick Gordon Guggisberg's administration to serve as a green buffer between Achimota School and the city of Accra. The other two initiatives, AUGIP and TGRP, are privately-initiated with either a total focus on green spaces or a component focus on greening within its scope. AUGIP is entirely focusing on green infrastructure within the Greater Accra Metropolitan Area while TGRP has only component parts of their projects devoted to greening.

The four (4) initiatives were first assessed on the dichotomous responses of 'Yes' or 'No' to see whether the green initiatives incorporate the principles of good governance in their greening efforts. Next, the extent of incorporation of the principles of good governance was assessed using information from the greening initiatives. Information sources included quarterly and annual reports, and other forms of information, published through the organizations' websites and other publications in mainstream media. Data were first analyzed to find out how well it fits into the good governance principles before further analysis on the extent of incorporation (Table 1). Data were further checked for completeness and consistency before it was assessed in urban greening framework (Buijs et al., 2019). Only data which were published by the initiatives and available through their websites and other domains were used.

4 Analysis of Findings

This section presents data analysis based on the selected initiatives. It characterizes the four urban greening initiatives, emphasizing their focus, governance types, membership, tenure arrangements, resources, and institutional formality in urban greening (Table 2). The section proceeds to analyze data and assess urban greening by the initiatives using the good governance principles.

4.1 *Analysis of Good Governance Principles in Urban Greening Initiatives*

4.1.1 *Legitimacy*

The first principle of good governance considered was legitimacy. Basically, legitimacy is the acceptance of decisions, actions, and outcomes of an organization (Lockwood et al., 2010). For an organization to be legitimate, it must be accepted by regulating stakeholders and the participant communities. There are various ways through which legitimacy was assessed in the governance of urban greening initiatives. First, the Green Ghana Project (GGP) got its legitimacy as a result of it being sanctioned by the elected president of Ghana, who asked all organizations and individuals to prioritize tree planting. Another form of legitimacy in GGP was the involvement of various organizations; ministries, public and private agencies, the traditional authorities, and the general public. Given the widely accepted and circulated nature with which people and agencies received the initiative, a certain level of acceptance, known as legitimacy, was achieved.

The traditional authority is a highly revered institution in Ghana, since traditional leaders hold the lands in trust for their communities. The participation of traditional leaders in the Green Ghana Project (GGP) was a way of leveraging support for sustainable urban greening. One of the most powerful chiefs in Ghana, Otumfuo Osei Tutu II sent a delegation to plant a tree as a way of showing his support for the Green Ghana Project. Many other chiefs followed this example by planting trees to mark the day. The massive participation of stakeholders and people with diverse interests signals the legitimacy of the GGP.

The legitimacy of Achimota Forest (AF) rests in the state, managed by the Ghana Forestry Commission and the Lands Commission. As a state-owned organization, AF is managed for and on behalf of the people of Ghana. The forest reserve which began with 494.95 hectares in 1930 has been constantly encroached, reducing it to 214 hectares in 2022. The legitimacy of the state to protect AF, however, came to contention in May 2022, following an executive instrument (E.I 144) by government

Table 2 Characteristics of selected case study sites

CASES	*Case 1: GGP*	*Case 2: AF*	*Case 3: AUGIP*	*Case 4: TGRP*
Scale	National	Regional	Regional	Catchment
Project focus	Integrated environmental management, including greening	Urban green space, providing ecological and other ecosystem services to Accra	Restoring urban green in the national capital city of Ghana, Accra	Integrated environmental management with green development projects
Governance type	Hierarchical governance, central government controls resources	Hierarchical; government department responsible for administration	Closed co-governance. Membership is private but cooperates with government actors	Open co-governance. Membership is opened to government and non-government actors
Organizational members	Government and its decentralized institutions and agencies, individuals	Government institutions such as lands commission and forestry commission	Local governments Friends' groups State agencies Private actors	Local governments, institutions, individuals, and friends
Others (non-member actors)	Traditional authority, civil society groups	National development planning commission	Landscape experts, local residents	UN (funding provision), schools, and resident groups
Land tenure	Public ownership and management (national scale)	Public ownership and management	Private ownership and management	Private ownership and management
Resources	Central government	Central government	Grants, both public and private	Sponsorship from government and non-government actors
Institutional formality	Government bodies, institutions, and the general public	Gazetted since 1930 to provide green buffer between Achimota School and the city of Accra	Network of local actors, interest groups, and philanthropist	Registered and incorporated association

which sought to declassify 41% of the forest to pave way for built up areas. Albeit E.I 144 is being implemented, environmentalist and civil society organizations have raised concerns about its ecological injury to the forest.

With regard to the other two initiatives, TGRP and AUGIP, a careful review of their past and ongoing projects reveals that the initiatives continue to benefit from continuous funding from their sponsors for implementing green projects. The continuous funding by the sponsors is in part due to the diligence, standardization, and trust held by the sponsors in the outcomes of previous and current projects. This indicates legitimacy since the sponsors and their operation communities have accepted their actions as contributing to environmental governance.

4.1.2 Transparency

Transparency is explained as availability of information, concerning the decision-making process and performance of an organization. The findings show that the various organizations have varying levels of transparency. For example, with the exception of the Green Ghana Project which does not have a visible transparency index for measuring its success. All other initiatives, Achimota Forest (AF), Accra Urban Green Infrastructure Project (AUGIP), and The Green Republic Project (TGRP), have some form of reportage or visibility for transparency. The transparency of AF rests in the Forestry Commission and the Lands Commission, the two government agencies which manage the forest. AUGIP and TGRP have websites, reporting formats, and media platforms through which previous projects, administration, and deliverables are published. Again, all the urban greening initiatives except the Green Ghana Project have been duly registered to practice.

In the instances, where there is no publicly available information on the decision-making processes of an initiative, the range of stakeholders involved in decision-making as well as how grants are managed and administered were assessed. For example, AUGIP and TGRP have formalized reporting systems with extension of invitation to third parties who are willing to attend discussions on their decision-making. Although AF is a state-owned property, it has incorporated third party such as Land Use and Spatial Planning Authority (LUSPA) and other stakeholders in its decision-making and governance to ensure transparency. The GGP, however, is a new initiative and does not have clear-cut decision-making bodies with visible transparency in decision-making. Nevertheless, being a state-led initiative, the transparency of the GGP is tied to the transparency of the initiating government. The implication is that, beyond making information publicly available, transparency may be assessed or gauged by the willingness of institutions to respond to issues about its governance structure, resource ownership, management, and use. This promotes public confidence and involvement in decisions toward urban greening.

4.1.3 Accountability

Accountability means who is responsible to who and for what task? This analysis focuses on financial accountability, despite the existence of many forms of accountability (Mulgan, 2000). Due to different endowments, the selected cases demonstrate accountability in different ways. Although all the four initiatives have some level of accountability, much efforts are needed to boost public confidence. Again, the different levels of organizational formality and governance influence the variations in approach to governance.

It was realized that the AUGIP and TGRP have both received state and/or international funding of a sort at one point in time. AF is managed by state agencies the Forestry Commission and the Lands Commission. The two agencies are accountable to the general public for managing the forest through the government. Again, the GGP is state-led and is being funded by the state under the auspices of the Ministry of Lands and Natural Resources. The implication is that, transparency is one of the major requirements for national and or international funding. Again, while AUGIP and TGRP provide reports on spending to their funding bodies, AF and GGP are audited as part of the audit of decentralized government agencies in accordance with relevant provisions of the 1992 constitution of Ghana. Again, self-funding from AUGIP and TGRP ensures the organizations provide statements on spending to its audience.

4.1.4 Inclusiveness

Inclusiveness means the level of involvement of stakeholders in the decision-making process of the organizations. Consequently, this study finds all four urban greening initiatives inclusive as all of them sought to promote the participation of interested stakeholders in their activities. Although the levels of participation vary as a result of the different spatial and or temporal scales of the projects, the initiatives were generally inclusive. The GGP for instance called on all Ghanaians irrespective of their political and/or ideological leaning to participate in the promotion of greenery by planting trees. AF has many stakeholders involved in its governance and administration including chiefs, government agencies, citizens, and private partners. The AUGIP and TGRP also encourage collaborations and participation of different interest groups in carrying out green activities. Despite these efforts, inclusivity in greening was assessed to be on the decline.

All the four initiatives were rather silent on encouraging gender balance in the participation of their programs and activities. Again, not until the executive instrument (E.I 144) was signed by government, many people did not know about the governance and management of AF. This smacks of exclusiveness in the governance and administration of the forest. TGRP has incorporated empowerment into its frameworks, which is important in ensuring comprehensive participation. Coffey et al. (2020) argues that there are different ways of assessing inclusiveness. This brings to light issues of who is responsible for planting trees, who supplies the trees, and who monitors the planting? These rather relevant questions have implications on how urban greening is perceived, approached, and practiced by the general public. GGP witnessed massive participation from all aspects of people including the political elites, traditional authorities, civil society organizations, and local governments. AUGIP is responsible to its local base as well as funders in promoting urban greening.

4.1.5 Fairness

Fairness measures absence of biases in sharing the benefits and cost of an intervention. It deals with consistency, absence of abuse, and respect for the opinions. All the four urban greening initiatives were found to comply with the principle of fairness, albeit at different scales. TGRP for instance undertakes some of its projects in educational and other facilities which affects the general public. It also acknowledges the role of the youth in responding to the call for climate protection through urban greening. To this regard, TGRP engages in projects that promote youth development, especially in the green economy. Put differently, urban greening has lots of social co-benefits. It regulates urban heat, serves as environmental amenity, enhances biodiversity, and promotes employment. GGP contributes to climate change adaptation through improved forest cover which regulates ecosystems of various kinds. AF provides a green buffer between the city of Accra and Achimota School. It regulates the microclimate and provides a range of ecosystem services to its surrounding communities. AUGIP has urban greening components and community-based initiatives. These functions comply with fairness because the costs and benefits of these interventions are widespread and unbiased to the general public.

Among the four initiatives, however, the ability to resolve the wider issues regarding equity in the delivery of benefits and costs arising from greening initiatives has been missing. Nevertheless, because most of these projects have public good components, the positive externalities in terms of amenity, biodiversity, and ecosystem protection are enjoyed by all regardless of their status in society. Also, conscious efforts have been made by AF, TGRP, and AUGIP in providing carbon sink, youth empowerment in greening efforts, and environmental awareness respectively.

4.1.6 Integration

Integration is understood as the ways by which urban greening initiatives are fused into other sectors and levels of decision-making. Integration could be vertical (other levels) or horizontal (other sectors) in decision-making. The four initiatives were found to have complied with the principle of integration, albeit there were variations in the nature of incorporations. AF, TGRP, and GGP have lots of vertical integration through collaborations with various levels of governments (including local, central government, ministries, departments, and agencies). GGP and TGRP also integrate well with the traditional authorities, institutions, and individuals at different levels. Horizontal linkages were also found to be strong with AUGIP in the form of collaboration with the Departments of Parks and Gardens, Urban Roads, Health, and Education sectors in promoting green development ideas especially among the youth.

However, the horizontal linkages were weaker than the vertical ones. Apart from TGRP and AF which had tangible integrations with education and socio-economic sectors, other benefits were intangible. AUGIP also has links with environmental sanitation and well-being which is encouraged. Although TGRP has done a lot of work in advocacy for environmental safety and has strategic links with government agencies and departments, general vertical collaborations were average. More advocacy is needed to protect the already existing forest cover, while planting new trees. It could be inferred that due to limited levels of environmental advocacy, precious ecosystems such as the Achimota Forest and Atiwa

Forest are being destroyed in attempts to pave way for built up areas and mineral exploration. It will, therefore, take more advocacy to resist subsequent efforts in protecting the forest cover. To this extend, only TGRP has programs focusing on environmental advocacy. This calls for policy innovations, aimed at mainstreaming efforts for green spaces development in urban planning frameworks (Hildén et al., 2014). The current green initiatives are not well integrated due to lack of policy to champion the development of urban green spaces.

4.1.7 Capability

Capability refers to the ability to initiate and carry out actions. Capacity is often fulfilled as a responsibility, although it could go beyond the immediate responsibilities. Findings revealed that the selected initiatives have demonstrated the capacity to act at some levels especially in pro-environmental programs. The initiation and promotion of greening for instance by the initiatives show commitment toward environmental safety.

Despite the common interest in the capacity to act, each of the four initiatives operate differently. The GGP for instance demonstrates the capacity to act by declaring "Green Ghana Day," a short-term initiative of the long-term "Green Ghana Project" (GGP) aimed at restoring Ghana's tree cover within the next five years by planting trees. It started with the planting of seven million trees in 2021 and 20 million trees in 2022.

AF is an ecological area which provides a variety of health, social co-benefits, and environmental functions for the sustainability of the urban ecosystem in the city of Accra. TGRP demonstrates capacity through training and retraining of individuals in volunteering and other soft skills in order to take up environmental management challenge. AUGIP, on the other hand, utilizes its local networks and professionally designed initiatives to improve the skills of its members while cooperating directly with local and central government agencies to improve environmental sanitation and green infrastructure in Accra. The four initiatives collaborate with researchers to produce and share knowledge in their areas of operation.

Fundamentally, GGP rallied diverse stakeholders to make efforts in order to increase the national carbon sink through tree planting. These add to the diverse ways of building capacity to enable reasonable adaptation to external shocks. Although all these initiatives contribute to climate change adaptation, their framing were not targeted as climate change interventions, hence limiting their impact. More importantly, the lack of urban greening policy affected the capacity of the initiatives to contribute to climate change adaptation. The resulting impact is the lack of focus for green development as the President of Ghana signed an Executive Instrument (E.I 144) effective May 1, 2022, to declassify some 146.30 hectares of the Achimota Forest (AF). Environmentalists have blamed this move, calling for the sanctity of the AF to be preserved as an ecological area.

4.1.8 Adaptability

Adaptability is the integration of relevant novel information into decision-making. It also deals with reflection of the performances of organizations in order to strategize for better outcomes. TGRP remains resolute to its core mandate of environmental safeguarding by increasing awareness and the need to plant more trees. It has also engaged the youth, educational institutions, and diverse stakeholders in the green revolution by initiating and planting trees.

The AF through a blend of environmental and social co-benefits has influenced the economy of Achimota area, while remaining committed to environmental safety and preservation of the forest. However, governmental irresponsibility has often led to the encroachment of parts of the forest, causing serious ecological injury. TGRP has engaged local communities, government agencies, and environmentalists through co-production of knowledge. It has also used its platform to call for actions to initiate environmental protection protocols to help adapt to the impact of climate change. The TGRP has been credited as its advocacy is probably responsible for the initiation of the Green Ghana Day by the government of Ghana. AUGIP has trained and mentored members in the acquisition of new skills and environmental awareness to take advantage of the emerging opportunities presented by climate change. Table 3 indicates the level of incorporation of good governance principles by greening initiatives.

5 Discussion of Findings

5.1 The Need for UGS Policy in Ghana

Urban greening is an indispensable tool for urban planners who seek to promote livable cities. Several cities in the developed countries such as New York, Chicago, Washington DC, and San Diego have clearly

Table 3 Summary of incorporation of good governance principles in greening initiatives

Good governance principle	*Green space initiative*			
	GGP	*AF*	*AUGIP*	*TGRP*
Legitimacy	Obtained from elected government and diverse stakeholder participation	Rests in the state through Forestry Commission and the Lands Commission	Continuous improvement and implementation of its funded projects	Funder approved legitimacy through continuous support of organization
Transparency	Does not have visible transparency index (transparency tied to sitting government)	Tied to the transparency of Forestry Commission and the Lands Commission	Highly transparent. Involves third party in decisions. Have formalized reporting systems	Transparent with existence of platforms for third party in decisions. Have formalized reporting system
Accountability	Lies in Ministry of Lands and Natural Resources, and Forestry Commission	Managed by Forestry Commission and Lands Commission in collaboration of other departments and sectors	Provides information on both national and international funding and disbursement of funds	Provides information on funding and disbursement of funds to funding bodies
Inclusiveness	Ghanaian public and private sector institutions, traditional authority, and individuals	Involves central and local government, LUSPA, agencies, and departments in decision-making	Have diverse stakeholder involvement in decision-making. Silent on gender balance in measurements	Incorporated empowerment for young people, but lacks gender balance in inclusivity

(continued)

Table 3 (continued)

Good governance principle	*Green space initiative*			
	GGP	*AF*	*AUGIP*	*TGRP*
Fairness	Wider coverage, cutting across different stakeholders but no focus on equity of environmental costs and gains	Provides a green buffer between the city of Accra and Achimota School. Increases carbon sink	Urban greening and community-based green infrastructure projects. No focus on equity	Projects involve youth development and the role that they play in greening projects. Equity initiatives missing
Integration	Mainly vertical integration through collaborations with government agencies and departments, and chiefs	Both vertical and horizontal integration through linkages with central and local government, departments, and agencies	Linkages with Urban Roads, Health, and Education sectors in promoting green development	Has vertical integration through collaborations with various levels of governments. Have advocacy projects
Capability	Green Ghana Day once every year for five years with the aim of restoring forest cover as climate change intervention	Provides a variety of environmental, health, social, and economic benefits to its area	Uses local networks and professionally designed initiatives to improve environmental issues	Capacity through training of youth in volunteering and environmental advocacy. Train youth for climate change actions
Adaptability	Initiate and motivate people to coordinate and plant trees	Environmental safety and social co-benefits through greening (forest)	Trained and mentored members in the acquisition of new skills relevant for environmental safety	Engaged the youth in planting trees. Their advocacy has led to the Green Ghana Project by Ghana government

articulate urban greening policies which guide the development and redevelopment of cities. In the developing countries, cities such as Kampala in Uganda and Durban (South Africa) have urban greening policies. In Ghana, however, there is currently no urban greening policy to guide the development of urban areas. As a result, urban greening initiatives in Ghana appear to be unstandardized and spontaneous. Greening initiatives, therefore, continue to suffer from standardization and consistency since actions are not backed by a thought-through urban greening policy.

Although many initiatives in Ghana contribute to the development of urban greening, the bigger issue of urban greening sustainability is at stake as lack of policy clouds the vision of green initiatives. The lack of policy on urban greening may also be accountable for the inability to use UGS to improve the adaptive capacity of cities to climate change. Among the eight principles of good governance, fairness, and inclusiveness were the hardest to comply with as many of the initiatives operate at the local level without recourse to equity and standardization. Accountability was also not clearly incorporated in some of the initiatives. Beyond this, the initiatives indicate intensions to integrate UGS in climate change adaptation. In particular, the TGRP is recognized for its role in leading actions for tree planting as a climate change intervention. However, the lack of policy innovation hampers its development since it cannot scale up the initiatives to other parts of the country. Policy direction may be useful in consolidating climate change adaptation through greening initiatives. These initiatives will better function if there is urban greening policy to direct the course of UGS development. The lack of a national UGS policy makes it difficult to pursue a common vision in this era of green revolution.

Urban greening must, therefore, be socially motivated and environmentally embracing in addressing both environmental sustainability and social co-benefits through effective policy framing. This corroborates with initiatives in Durban city, South Africa where both environmental and social co-benefits have been achieved through urban greening (eThekwini Municipality and ICLEI Africa Secretariat, 2007; Shih, 2017). For instance, the need to integrate different stakeholders in greening initiatives (Buijs et al., 2016); consideration of the drivers and spatial scales for incorporation of UGS (García et al., 2018); political actors, urban planners, and traditional authorities (Amoako & Adom-Asamoah, 2019); and the social benefits of UGS (García et al., 2018; Kitha & Lyth, 2011) must be environmentally relevant and socially all-embracing.

Reflecting on good governance principles outlines the role of urban greening policy in framing urban green spaces as climate change interventions. In particular, good governance promotes the co-design, co-production, and co-governance of urban greening through interactions with both the public and private agencies (Kitha & Lyth, 2011; Lockwood et al., 2010). Lastly, lessons can be drawn from San Diego, New York, Chicago, Kampala, and Durban cities in framing UGS as climate change interventions for optimum benefits.

5.2 *The Potentials of Urban Greening Initiatives for Climate Change Adaptation*

Each of the four urban greening initiatives has goals targeting climate change as documents frequently made reference to climate change adaptation. This supports the calls in UGS literature for innovative means to addressing climate impacts (Coffey et al., 2020). However, historical information indicates that initial purpose for establishing Achimota Forest (AF) was not for climate change adaptation. This is not surprising because climate change, perhaps, was not a topical issue at the time.

Nevertheless, current discussions about the forest have linked its relevance to the ability to sequester carbon emissions, provide the largest green cover to the city of Accra, and serve as nature-based stormwater drains, preventing flooding in the area (Puplampu & Boafo, 2021). The other three initiatives started with clear focus on climate change from inception, albeit with deferring approaches to mitigation and adaptation. For example, the primary aim of the Green Ghana Project (GGP) was to restore green cover and encourage tree planting among the youth to tackle climate change. This project planted a total of 27 million trees in 2021 and 2022 in efforts to recover the forest. With 80% survival rate of the planted trees, Ghana's forest cover will receive a significant boost, leading to improved carbon sequestration. This initiative has potentials to contribute to reduction in urban heat due to regulation of temperatures and control flooding through a boost in carbon sink (Sturiale, 2019). Projects from AUGIP and TGRP have the potentials to retain significant amount of rainfall by increasing green cover. This reduces urban heat waves and serves as stormwater drains, leading to a reduction in risks of flooding. The initiatives also contribute to improved air and water quality (Reckien et al., 2014) and sustainable micro-environment.

On the other hand, findings from the Accra Urban Green Infrastructure Project (AUGIP) and The Green Republic Project (TGRP) show inclusiveness in the creation of large pristine UGS for public relaxation and enjoyment and strong advocacy for green spaces. AUGIP trains specialists to impact urban areas by nurturing greenery in urban areas. This is important in the wake of calls for a change from gray to green infrastructure in response to the impacts of climate change (Caprioli et al., 2021). Again, AUGIP partners environmental projects and research aimed at knowledge creation and distribution among environmentalists. Mabon and Shih (2021) assert that sound environmental governance practices are anchored on the ability to create and share data in critical UGS areas such as green roofs and other forms of greening as nature-based adaptation to climate change. TGRP is an environmental advocacy group and is partly responsible for the Ghana government's Green Ghana Project. TGRP engages the youth on creating green spaces and trains people on environmental advocacy in response to climate change impacts. Environmental advocacy is needed in order to protect the gains made in urban greening as Cobbinah and N-yanbini (2019) indicate that lack of environmental advocacy derails the gains made in climate change adaptation. This is particularly important in preserving UGS as low-cost and nature-based adaptation to climate change (Kitha & Lyth, 2011).

All the four urban greening initiatives serve functions beyond environmental sector to include economic and social co-benefits. For example, Achimota Forest (AF) attracts 20,000 visitors per annum and generates US$60,000 annually in revenue while serving as meeting grounds for societies and religious groups. The forest, which also incorporates an eco-park, provides employment while serving as adaptive responses to climate change. Similarly, GGP, AUGIP, and TGRP provide opportunities for employment and youth participation in urban greening. However, the Achimota Forest has been threatened by the latest declassification of 41% of its reserves by the Ghanaian president in May 2022 through executive instrument (E.I. 144). Despite resistance by environmentalists about the ecological injury caused by E.I. 144, the president went ahead to declassify parts of the forest. The constant encroachment of the forest which has reduced its size from 494.95 hectares in 1930 to 214 hectares in 2022 presents worrying trends for its sustainability. Despite its ambitious adventure by GGP, AF, AUGIP, and TGRP, lack of urban greening policy has thwarted the efforts in promoting green development. This lack of policy direction creates confusing for greening efforts as UGS across all the four cases lacked long-term visions. Even where these visions exist, translating them into actions requires expert coordination.

6 Conclusion and Recommendations

Fundamentally, this chapter explores the governance of urban greening initiatives through the application of good governance principles from natural resource management perspective. It contributes to understanding the complexity of issues involved in the governance of green spaces. It also contributes to understanding the integration of urban greening governance as a climate change intervention. In particular, this study contributes to understanding how urban greening initiatives could go beyond environmental benefits to include social and economic co-benefits.

It also demonstrates the application of a lucid set of criteria through good governance principles (Lockwood et al., 2010) to understand multifaceted ways of improving upon the governance of green spaces. Different perspectives of governance have been revealed through the application of the principles of good governance in urban greening. This chapter finds that three principles (inclusivity, fairness, and accountability) were challenging to achieve due to the lack of policy to guide urban greening in Ghana. The chapter calls for dynamic approaches to tackle critical issues of social equity, environmental amenity, and biodiversity protection through the initiation of urban green spaces policy.

The chapter draws on good governance principles by emphasizing that while imploring international standards to the governance of urban greening, local benefits and the role of green spaces in climate change adaptation should be prioritized. This is important not only for the governance of urban greening but also due to its contribution to climate change adaptation. This is particularly important for developing countries as urban greening proves to be a cost-effective means of climate change adaptation.

The chapter acknowledges that increased advocacy and focused environmental policies have larger environmental and social co-benefits. It recommends the integration of social needs in urban greening policies to leverage the many social, economic, and environmental co-benefits of greening. It also calls for advocacy by urban planners in attracting the attention of policymakers and the political elites to prioritize urban greening development. The chapter proffers the formulation of sustainable urban greening policy as a pathway to climate change adaptation.

References

Amoako, C., & Adom-Asamoah, G. (2019). From the seat of a traditional Kingdom to a Garden city: The socio-spatial politics of managing green areas in Kumasi, Ghana. *African Geographical Review, 38*(4), 310–325.

Arnouts, R., van der Zouwen, M., & Arts, A. (2012). Analysing governance modes and shifts—Governance arrangements in Dutch nature policy. *Forest Policy and Economics, 16*, 43–50.

Asibey, M. O., & Cobbinah, P. B. (2022). The evidence for climate change on our planet. *The Palgrave handbook of global sustainability* (pp. 1–16). Cham: Springer International Publishing.

Buijs, A., Hansen, R., Van der Jagt, S., Ambrose-Oji, B., Elands, B., Lorance Rall, E., Mattijssen, T., et al. (2019). Mosaic governance for urban green infrastructure: Upscaling active citizenship from a local government perspective. *Urban Forestry and Urban Greening, 40*, 53–62.

Buijs, A. E., Mattijssen, T. J., Van der Jagt, A. P., Ambrose-Oji, B., Andersson, E., Elands, B. H., & Steen Møller, M. (2016). Active citizenship for urban green infrastructure: Fostering the diversity and dynamics of citizen contributions through mosaic governance. *Current Opinion in Environmental Sustainability, 22*, 1–6.

Caprioli, C., Oppio, A., Baldassarre, R., Grassi, R., & Dell'Ovo, M. (2021). A multidimensional assessment of ecosystem services: From grey to green infrastructure. *International Conference on Computational Science and Its Applications* (pp. 569–581). Springer.

Cobbinah, P. B., Asibey, M. O., Opoku-Gyamfi, M., & Peprah, C. (2019). Urban planning and climate change in Ghana. *Journal of Urban Management, 2*, 261–271.

Cobbinah, P. B., & N-yanbini, N. N. (2019). Climate change adaptation in Urban Ghana: The spatial planning dimension. In P. Cobbinah & M. Addaney (Eds.), The geography of climate change adaptation in Urban Africa. Palgrave Macmillan, Cham. https://doi.org/10.1007/978-3-030-04873-0_15

Coffey, B., Bush, J., Mumaw, L., de Kleyn, L., Furlong, C., & Cretney, R. (2020). Towards good governance of urban greening: Insights from four initiatives in Melbourne, Australia. *Australian Geographer, 51*(2), 189–204.

Diko, S. K., & Hollstein, L. M. (2021). Towards an alternative interpretation of the socio-cultural dimensions of urban greenspace planning in the Global South: Evidence from the Kumasi Metropolis of Ghana dimensions of urban greenspace planning in the Global South. *Journal of Urban Affairs, 00*(00), 1–24.

Escobedo, F. J., Kroeger, T., & Wagner, J. E. (2011). Urban forests and pollution mitigation: Analyzing ecosystem services and disservices. *Environmental Pollution, 159*(8–9), 2078–2087.
eThekwini Municipality and ICLEI Africa Secretariat. (2007). Durban biodiversity report 2007, Local Action for Biodiversity (LAB) Report, Durban, South Africa.
Etingoff, K. (2016). *Urban ecology: Strategies for green infrastructure and land use*. Apple Academic Press Toronto.
García, F., Solecki, W. D., & Ribalaygua, C. (2018). Climate change adaptation in Europe and the United States: A comparative approach to urban green spaces in Bilbao and New York City. *Land Use Policy, 79*(August), 164–173.
Gill, S. E., Handley, J. F., Ennos, A. R., & Pauleit, S. (2007). Adapting cities for climate change: The role of the green infrastructure. *Built Environment, 2007*(1), 115–133.
Govindarajulu, D. (2014). Urban climate urban green space planning for climate adaptation in Indian cities. *Urban Climate, 10*, 35–41.
Gupta, K., Kumar, P., Pathan, S. K., & Sharma, K. P. (2012). Urban Neighborhood Green Index—A measure of green spaces in urban areas. *Landscape and Urban Planning, 105*(3), 325–335.
Hildén, M., Jordan, A., & Rayner, T. (2014). Climate policy innovation: Developing an evaluation perspective. *Environmental Politics, 23*(5), 884–905.
IPCC. (2019). IPCC Special report on climate change, desertification, land degradation, sustainable land management, food security, and greenhouse gas fluxes in terrestrial ecosystems. *IPCC SRCCL*.
Kendal, D., Lee, K., Ramalho, C., Bowen, K., & Bush, J. (2016). *Benefits of urban greenspace in the Australian context: A synthesis review for the clean air and Urban landscape hub*. University of Melbourne.
Kitha, J., & Lyth, A. (2011). Urban wildscapes and green spaces in Mombasa and their potential contribution to climate change adaptation and mitigation. *Environment and Urbanization, 23*(1), 251–265.
Lawrence, A., De Vreese, R., Johnston, M., Konijnendijk van den Bosch, C. C., & Sanesi, G. (2013). Urban forest governance: Towards a framework for comparing approaches. *Urban Forestry and Urban Greening, 12*(4), 464–473.
Lemos, C., & Agrawal, A. (2006). Environmental governance. *Annual Review of Environment and Resources, 31*, 297–325.
Lindley, S., Pauleit, S., Yeshitela, K., Cilliers, S., & Shackleton, C. (2018). Landscape and urban planning rethinking urban green infrastructure and ecosystem services from the perspective of sub-Saharan African cities. *Landscape and Urban Planning, 180*, 328–338.
Lockwood, M. (2010). Good governance for terrestrial protected areas: A framework, principles and performance outcomes. *Journal of Environmental Management, 91*(3), 754–766.

Lockwood, M., Davidson, J., Curtis, A., Stratford, E., & Griffith, R. (2010). Governance principles for natural resource management. *Society and Natural Resources, 23*(10), 986–1001.

Mabon, L., & Shih, W. Y. (2021). Urban greenspace as a climate change adaptation strategy for subtropical Asian cities: A comparative study across cities in three countries. *Global Environmental Change, 68*, 102248. https://doi.org/10.1016/j.gloenvcha.2021.102248

Mayrand, F., Clergeau, P., Vergnes, A., & Madre, F. (2018). Vertical greening systems as habitat for biodiversity. In G. Pereez & K. Perini (Eds.), *Nature-based strategies for urban and building sustainability* (pp. 227–237). Elsevier.

Mensah, C. A., Andres, L., Beazley, M., & Roji, A. (2015). Managing urban green spaces in Africa: A collaborative governance approach. In Z. U. Arif, N. Rajput, & A. Şener (Eds.), *Advances in resource management and consumer sciences*. ORIC Publication.

Mulgan, R. (2000). 'Accountability': An ever-expanding concept? *Public Administration, 78*(3), 555–573.

Pearlmutter, D., Calfapietra, C., Samson, R., O'Brien, L., Ostoić, S. K., Sanesi, G., & del Amo, R. A. (2017). The urban forest. In *Cultivating green infrastructure for people and the environment* (Vol. 7). Springer.

Puplampu, D. A., & Boafo, Y. A. (2021). Exploring the impacts of urban expansion on green spaces availability and delivery of ecosystem services in the Accra metropolis. *Environmental Challenges, 5*, 100283.

Reckien, D., Flacke, J., Dawson, R. J., Heidrich, O., Olazabal, M., Foley, A., Hamann, J. J. P., Orru, H., Salvia, M., De Gregorio Hurtado, S., Geneletti, D., & Pietrapertosa, F. (2014). Climate change response in Europe: What's the reality? Analysis of adaptation and mitigation plans from 200 urban areas in 11 countries. *Climate Change, 122*(1–2), 331–340.

Robberts, D. C. (1994). The design of an urban open-space network for the City of Durban (South Africa). *Environmental Conservation, 21*(1), 11–17.

Safransky, S. (2014). Greening the urban frontier: Race, property, and resettlement in Detroit. *Geoforum: Journal of Physical, Human, and Regional Geosciences, 56*, 237–248.

Schultz, L., Folke, C., Österblom, H., & Olsson, P. (2015). Adaptive governance, ecosystem management, and natural capital. *Proceedings of the National Academy of Sciences, 112*(24), 7369–7374.

Shih, W. (2017, January). Greenspace planning for climate co-benefits in Durban, South Africa. *ResearchGate*.

Sturiale, L. (2019). The role of green infrastructures in urban planning for climate change adaptation. *Climate, 7*, 119. www.mdpi.com/journal/climate

United Nations. (2020). *Sustainable development goal 11*. Source: un.org/sustainabledevelopment/wp-content/uploads/2019/07/11_Why-It-Matters-2020.pdf. Accessed on 16 October 2022.

United Nations. (2018). *Revision of the world urbanization prospects*. United Nations.

van der Jagt, A. P. N., & Lawrence, A. (2019). Local government and urban forest governance: Insights from Scotland. *Scandinavian Journal of Forest Research, 34*(1), 53–66.

Wendel, H. E. W., Zarger, R. K., & Mihelcic, J. R. (2012). Accessibility and usability: Green space preferences, perceptions, and barriers in a rapidly urbanizing city in Latin America. *Landscape and Urban Planning, 107*(3), 272–282.

Yin, R. K. (2014). *Case study research: Design and methods* (5th ed.). Sage.

CHAPTER 9

Carbon Ownership and Issues of Resource Governance in Ethiopia

Zenebe Mekonnen

1 Introduction

The IPCC report 2022 shows that, particularly in Africa, climate change has already had increasing adverse impacts on human systems, including water security and food production, health and well-being, and cities, settlements and infrastructure (IPCC, 2022). To adapt to these challenges, it needs nations to develop new strategies. REDD + , which is connected to climate resilience, climate adaptation and policy innovations, is one of the strategies. First, *climate resilience* refers to the ability to self-organize, ability to buffer disturbance and capacity for learning and adapting that enable social or ecological systems to bounce back after a shock (Tompkins et al., 2005). It is worthwhile that understanding climate risks and their context enables effective planning for resilience. Second, *climate adaptation*, as one aspects of climate resilience, refers to making assets and people less vulnerable to the impacts of climate

Z. Mekonnen (✉)
Ethiopian Forestry Development, Addis Ababa, Ethiopia
e-mail: zenebemg2014@gmail.com

M. Addaney et al. (eds.), *Climate Change in Africa*,
https://doi.org/10.1007/978-3-031-30050-9_9

change (C2ES, 2019). Third, *policy innovations*-REDD + is one of the policy innovations, an invention of emission trading (Jordan & Huiteman, 2014) made to reduce emissions from deforestation and forest degradation through forest conservation and management so as to benefit the community through carbon financing by bringing on board the non-valued- and valued-carbon.

The climate resilient green economy (CRGE) strategy is the core development agenda of Ethiopia. Embedded in the CRGE program is Ethiopia's REDD + program that aims at contributing to the emission reduction goals of the forestry sector. Ethiopia has been implementing REDD + readiness activities since 2012 followed by significant progress in developing the National REDD + Strategy (NRS); National Forest Monitoring System for the Measurement, Reporting and Verification (MRV) of REDD + ; Forest Reference Level (FRL); and Safeguard Information System (SIS) (Andoh & Lee, 2018; Bekele et al., 2018; MEFCCC, 2018). Regardless of these efforts, the divergence between whether benefit should be shared according to efforts made to reduce deforestation and forest degradation or being based on property rights has created doubt between stakeholders and challenged the implementation of REDD + (Birhan et al., 2021; Duker et al., 2019; Pham et al., 2021). Key issue challenging the effectiveness and success of Ethiopia's REDD + is carbon rights definitions.

The question of who has the right to carbon is the question of who owns the land and forests. Uncertainty around forest use rights is a major cause of deforestation given that insecurity of land tenure provides little incentive for sustainable management and conservation of forested land (Alemie & Amsalu, 2020; Bekele, 2003). As a prerequisite for creating forest carbon, clarifying the use rights attached to forests has been identified as an important part of Ethiopia's carbon use strategy (Hedges, 2010; MEFCC, 2018). The *Constitution of Ethiopia (FDRE,* 1994*)* vests the ownership of land and natural resources in the State. This establishes a precondition for centralized forest management and tenure arrangements. If forest carbon is defined as a natural resource, it could be argued that forest carbon rights are also vested in the State. However, the *Constitution of Ethiopia (FDRE,* 1994*)* also permits every Ethiopian the right to own "private property," where private property is either a tangible or intangible product flowing from the work of individuals, collectives or communities. If an alternative view of forest carbon defines it as something "produced" by the communities that manage a carbon project and/or the private operators who invest in projects, the forest carbon rights might be considered to be private property—even though there is a presumption of State ownership over natural resources. On the other hand, the Forest Development, Conservation and Utilization Proclamation No. 542/2007 of article 3 had set private and state forest ownerships (FDRE, 2007). Under this Proclamation, article two, sub-articles 6 and 9, describes: "*state forest* means any protected or productive forest, which is under the ownership of the Federal Government or a Regional State; and *private forest* means a forest other than state forest developed by any private person and includes a forest developed by members of a peasant association or by an association organized by private individuals, investors and governmental and non-governmental organizations." In this case, the Proclamation has ignored community forest ownership including participatory forest management (PFM). Also from the definition of private forest, the power given to the private ownership seems to imply partial empowerment. Indeed, Proclamation 1065/2018 has categorized forest ownership into state, private, community and association forests (FDRE, 2018). Legal provision for carbon ownership is more complex because the carbon issue includes more complex international conventions and agreements in addition to national policy issues (Loft et al., 2017; Pham et al., 2021).

Although there are several challenges, there are efforts by the Ethiopian government toward REDD + implementation and carbon emission reductions (Bekele et al., 2015). The Ethiopian Government has given delegation of authority of governing forests to regional governments which enable them to take a lead role in defining an approach to forest carbon rights. These need to be addressed in order for carbon rights to be both established and appropriately shared. Granted, local forest resource rights in Ethiopian are strong "on paper" while weak "in practice" (Cotula & Mayers, 2009). Nevertheless, through provision of more transparent forest information systems, REDD + along with robust national forest strategy has the potential to improve forest governance and rights (Andoh & Lee, 2018; Angelsen, 2008) mostly done through participatory forest management (PFM). The objective of the paper is to deal with forest and carbon governance, carbon ownership, forest tenure arrangements, forest property rights and institutions, resource use conflicts and their management and forest carbon benefit sharing, and how all these affects REDD + implementation in Ethiopia based on the review of different literatures.

2 Forest Biomass Carbon in Ethiopia

To place a value on the carbon-bearing potential of any forested area, there is a need to accurately estimate how much carbon is being stored there. New technologies such as satellite imaging and computer modeling are making the measurement of carbon stocks both swift and accurate (CIFOR, 2008; Hairiah et al., 2020; Wodajo et al., 2020). A transparent system to measure and verify emission reductions now looks feasible. In Ethiopia, there are carbon initiatives in order to give value for the carbon in the forest. But there is variation in forest carbon stock quantity assessment in the country as indicated by different sources. The most comprehensive and detailed forest carbon stock quantification results in the country done so far include 153 million metric tonnes (MMT) of Carbon (Houghton, 1999), 2764 MMT (WBISPP, 2005), 867 MMT (Gibbs & Brown, 2007a), 2500 MMT (Nune et al., 2009), 593 MMT (FAO, 2010a) and 974 MMT (FAO, 2020). Indeed, carbon stock in forests is dynamic, the variation shown by these studies is very wide that might be inflected by data quality and assessment tools (Bosa et al., 2019). Data quality and the use of efficient assessment tools are one of the key pre-requisites to undertake REDD + strategies. The study by Woody Biomass Inventory and Strategic Planning Project of the Ministry of Agriculture (WBISPP, 2004) shows that most of the carbon stocks were concentrated in the then Southern Nations, Nationalities and people's Regional state followed by Oromia and Amhara Regional States (Fig. 1).

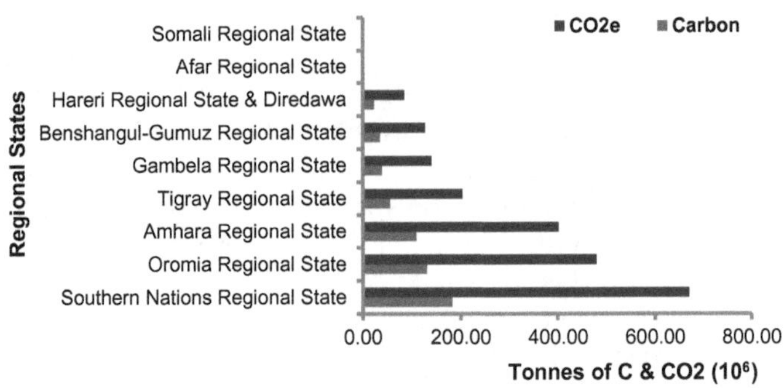

Fig. 1 Estimated Carbon stocks in Ethiopia's forests by regions (WBISPP 2004)

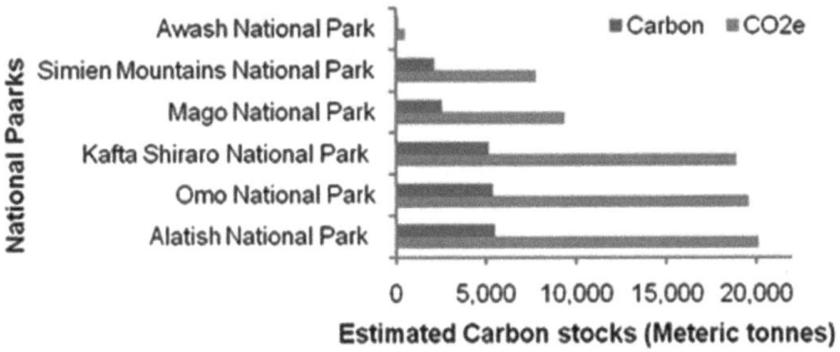

Fig. 2 Carbon stock estimates for six National Parks in Ethiopia (Vreugdenhil et al., 2011)

Another study "Carbon Baseline and Payments for Carbon Environmental Services from Protected Areas in Ethiopia" by World Institute for Conservation and Environment and the Manaaki-Whenua Land Care Research Institute on behalf of the Ethiopian Wildlife Conservation Authority has shown that there is considerable quantity of carbon stock in woody biomass (Fig. 2). However, in all of these studies, the question of who owns these carbon stocks is hardly underscored.

3 Forest Carbon Ownership and Governance

Governance refers to who makes decisions and how decisions are made, from national to local scale, including formal and informal institutions and rules, power relations and practices of decision-making (Fig. 3).

It also refers to the kinds of decisions that are made and whether they are clear, consistent and comprehensive. Hence, good forest governance means decisions are fair, transparent and just, rights are respected, laws and rules are enforced equitably, decision-makers are accountable, and decisions are made based on the analysis of what is good for people and forests in general and not personal interest (Larson & Petkova, 2011; Yami & Mekuria, 2022). Carbon governance refers to "influencing and conducting of the policy and affairs of institutions that aim to minimize the amount of carbon dioxide released to the atmosphere or maximize the amount of carbon stored stably away from the atmosphere. It might take the form of policy instruments such as taxes, incentives, regulations, subsidies, market-based mechanisms, voluntary measures, or new societal norms" (Dilling, 2007, p. 29).

Fig. 3 The three pillars of good forest and carbon governance (Larson & Petkova, 2011)

Achieving better forest governance for REDD + in Ethiopia, where tenure is not clearly defined, is an ambitious agenda that covers more than the institutional architecture for administering funds and monitoring results (CIFOR, 2010; ADF 2010; Gizachew et al., 2017). Indeed, good forest governance involves a range of issues to achieve better REED + implementation. These include the following (ADF 2010; CIFOR, 2010; USAID, 2012a; Gizachew et al., 2017):

- Recognition and enforcement of forest, land or carbon rights, including those of indigenous groups.
- Inclusion, participation and influence of forest communities and civil society in policy processes, institutional arrangements and setting management priorities.
- Effective, transparent and accountable forest management practices, and third party verification and scientific advisory groups to provide technical input;
- Effective, fair and proportionate enforcement and sanctions for illegal activities (including corruption and illegal logging).
- Responsive decentralized government, coordination between local and higher levels of government, inter-sectoral coordination, checks and balances to prevent domination by any one level or branch of government.
- Mechanisms for managing conflict effectively.

- Coherent actions built on accountable, transparent and participatory systems of governance that is key requirement in climate change management including carbon uses.
- Good governance and capable institutions allow users to participate more fully in decision-making processes and to articulate the specific needs and demands of a diverse society.
- In resource ownership and governance including carbon ownership, it is critically important to address the issues of inequality, fairness and opportunity for all members of resource users;
- To effect carbon ownership and carbon market, it requires an effective rule of law to prohibit illegal activities;
- Governance weakness for forest decline includes concentration of ownership to the state, weak or non-existent ownership and land tenure arrangements, illegal activities and corruption, etc.

As emphasized by CIFOR (2005), resource ownership issues that must be taken into consideration include: (i) conflict over resource use; (ii) lack of integration of indigenous knowledge and modern conservation methods; (iii) inadequate capacity of locals to manage resources; (iv) land tenure; and (v) lack of clear land use policy. Good governance is participatory (Fig. 4), consensus oriented, accountable, transparent, responsive, effective and efficient, equitable and inclusive and follows the rule of law. It ensures that corruption is minimized, the views of minorities are taken into account and that the voices of the most vulnerable in society are heard in decision-making. It is also responsive to the present and future needs of society (WRI, 2002; Lemenih et al., 2015; Rijal et al., 2021).

3.1 Tenure Arrangements

Land tenure refers to the full set of institutions and policies that determine how the land and its resources are accessed, who can hold and use these resources, for how long, and under what conditions. Land tenure, therefore, includes the broader collection of rights associated with the land and the institutions that uphold those rights (CIFOR, 2000, 2009, 2010; UNCCD 2017). Land and forest tenure is a central issue of concern for future carbon use policies and measures at country level. This is because tenure regimes define rights over forest resources and as such they determine who should be held responsible for making decisions on forest management and land use and who should be held responsible for losses

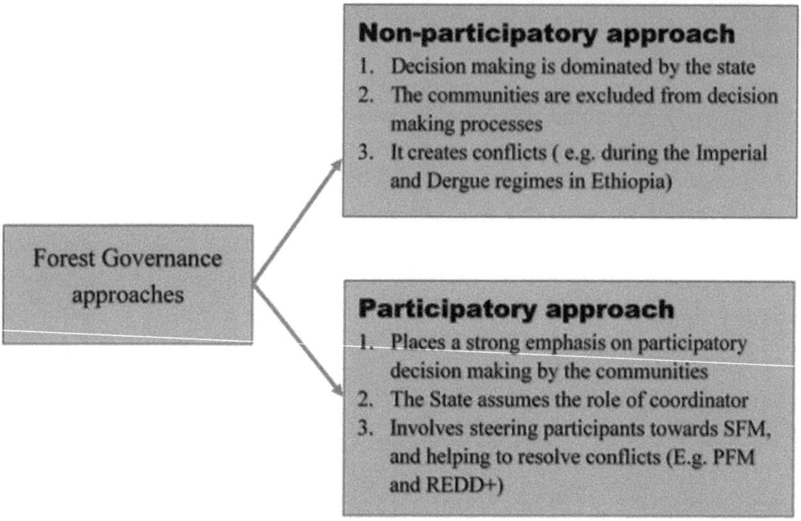

Fig. 4 Types of governance approaches and their implications (Lemenih et al., 2015)

or gains in forest carbon (Larson, 2012; Sunderlin et al., 2009). Forest tenure regimes also determine who can claim ownership and access to ecosystem services and their benefit streams, and these regimes will thus critically mediate the ability of carbon use policies and measures such as REDD + to achieve their outcomes (Corbera et al., 2011; Cotula & Mayers, 2009). Addressing tenure along with REDD + strategies can help strengthen forest governance and improve prospects for forest conservation but, not always with zero negative outcomes (Robinson et al., 2011).

To claim carbon ownership, an individual or group need to have a well-defined land use, for instance, a forest that can secure carbon sequestration and storage in order to be purchased by a voluntary carbon buyer. The State-policy-conflict-resource-institution nexus affects and determines how forest resources including carbon should be governed (Fig. 5). The Rural Land Proclamation 1997 of Ethiopia has given the authority for land administration to the regional governments and regional instruments must adhere to environmental law. But not all regions have enacted local laws. This indicates that tenure in Ethiopia is fragmented; customary tenures have been diminished and statutory tenure dominates and forms the basis of carbon rights (Hedges, 2010). Many laws in Ethiopia, including the Forest Development Proclamation 2007, deal with community consultation and participation but they do not reflect the extent to which they are translated into practice. Rather they give less emphasis regarding ownership and control of resources which in turn leads to undesired power disparity between communities and the State resulting in conflicts.

Literatures (*e.g.*, Hanna et al., 1996; Meinzen-Dick & Mwangi, 2008; Robinson et al., 2011) show that tenure systems, which affects carbon rights, can be categorized into four:

I. *Open access systems* at which access to natural resources is unregulated and open to everyone, and where it is difficult, costly, or almost impossible to establish rules of exclusion and regulation across resource users. This system is the common practice in Ethiopia in which forests are open access to everybody and become deforested due to overexploitation.
II. *State and public property regimes* where the state is the only institution with the legitimacy to vest access rights and management quotas over the resource to other users. Usually, the general public has equal rights to the resource and the State has coercive

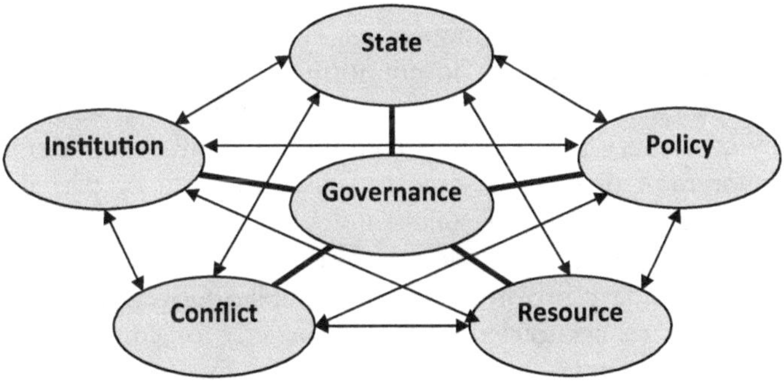

Fig. 5 Factors that influence resource governance and user rights (This woven interaction shows the impact of one on the other: for instance, conflict may arise due to resource scarcity, lack of good governance, institutional inefficiency, State interference, poor policy implementation)

powers of enforcement. Community rights are limited to procedural issues that rely on State authorities to realize those rights (Hedges, 2010). The government can establish regulations for sustainable resource use, but such regulations can be extremely costly to monitor and, as a result, become ineffective. In fact, public property is often unsuccessful in ensuring exclusion, and informal access to resources prevails. In many others instances, however, public property can de facto be used by individuals, organizations and/or communities who may hold long-term access and withdrawal rights over specific resources.

III. *Private property regimes* refer to situations in which individuals and families hold full rights over land and rely on state-based political and legal institutions to recognize and enforce their property claims. Private property holders have the right to exclude others from resources but the legitimacy of such rights would determine the costs of exclusion. Furthermore, property, particularly in forests, is often subject to regulations that in practice constrain how owners can manage their resources.

IV. *Common property regimes* where a group of resource users who share collective ownership over a territory, or over a single environmental resource at which the users share rights of access to and management of natural resources and rely on both community and state-based authorities to assert their claims, establish management rules and exclude outsiders, while the state retains alienation rights. Many traditional and indigenous rural communities in Ethiopia manage their resources in common but their—bundles of rights over such resources can be socially differentiated and regulated by customary practices and community institutions. Members of a common property regime can also hold full or partial private property rights over farming and grazing lands, which in some cases may be transferable to third parties, depending on legal and customary provisions. The Ethiopian Government has been reluctant to acknowledge the importance of common property and local systems of management. The theory of the "Tragedy of the Commons" is often used to justify government taking over common pool resources, on the grounds that local people do not have the expertise to manage them (Hardin, 1968). In Ethiopia, the absence of clear indicators in collective lands such as forests, rangelands and woodlands often results in these areas being classified as unproductive or unused (HLPE, 2011).

3.2 Property Rights and Institutions

Forest property rights (which also encompass carbon rights) in Ethiopia remained politically prompted, unjust, ill-defined and insecure at which the State failed to regulate the utilization of private and State forests (Ayana et al., 2013; Bekele, 2003; Bekele et al., 2015). The repercussion of this situation is: (i) overutilization and widespread deforestation, (ii) creation of conflicts among resource users and in turn these conflicts have consumed the forest resource and become incompatible with forestry development in the country. The forest property types in Ethiopia vary from regime to regime. During Imperial regime, there were state, private and community forests; Military regime, there were state and community forests; Ethiopian People's Revolutionary Democratic Front (EPRDF) regime, there were state; and private forests, current regime, there are state, private, community and association forests (Ayana et al., 2013; Bekele, 2003; FDRE, 2007, 2018).

Property rights are the products of rules, as mediated by formal and informal institutions, which liberate and constrain human action (Corbera et al., 2011; Naughton-Treves & Wendland, 2013). They liberate by authorizing an individual or collective to undertake particular actions, while at the same time constraining someone else to observe this right. While the full set of property rights can be held by public, private or community entities, it is often the case that an entity holds only a subset of the bundles of property rights (use, decision-making and alienation rights). In Ethiopia, particularly in pervious regimes, like in other African countries, the state typically holds administration rights to forests. However, since 2000 there is an increase in the formal devolution of use and decision-making rights, and in some cases alienation rights to individuals and community groups in the form of PFM (Ameha et al., 2014; Loft et al., 2015). Chilimo, Adaba-Dodola, Yabelo and Bonga forests are exemplary forests where PFM are well established in Ethiopia. In some cases, informal rights are more secure than formal rights in many forested areas of Ethiopia because the effective presence of the state to enforce formal rights is weak. (Fig. 6)

In Ethiopia, property rights to resources are weak. It is much weaker for carbon. As a result, competition to acquire them can be wasteful and

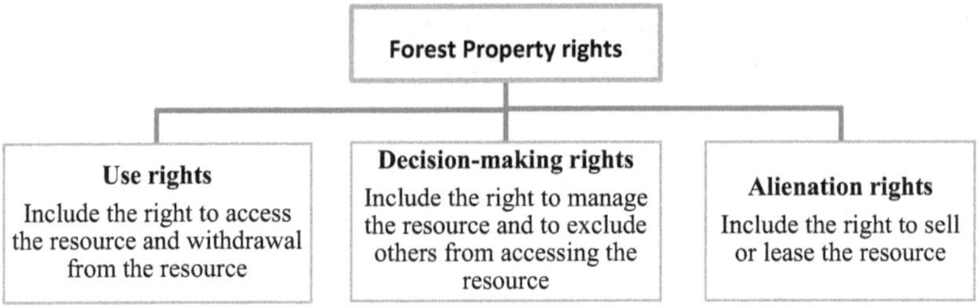

Fig. 6 Forms of resource property rights

be characterized by rent-seeking and violent conflict. The link to political systems comes about because ownership claims are most likely to be weak or ambiguous where the rule of law is not well established. Political institutions affect property rights and property rights in turn affect resource use. Political institutions also determine the State's ability to cooperate, coordinate and coerce. The role of forests in addressing the adverse effects of climate change and the need to strengthen forest sector research are among the premises set in Forest Development, Conservation and Utilization Proclamation #1065/2018. However, those institutions that have been obliged to effect those premises are inefficient and ineffective, and there is institutional instability in the forest sector. In the absent of appropriate political institutions, property rights may not provide incentives for efficient resource use as well as carbon marketing through the REDD + strategies (Deacon & Mueller, 2006).

3.3 *Resource Use Conflicts and Their Management*

Forest use conflicts are disagreements and disputes over access to, and control and use of forest resources. Conflicts often emerge because people have different uses for forests, or want to manage them in different ways, or existence of conflicts of interest between user groups (FAO, 2000). Understanding that conflict is a common feature of any resource use system and having proper approaches to manage and resolve them (Table 1) is a prerequisite for sustainable resource management that is participatory and equitable. The study (Mery et al., 2005) highlighted, conflicts may also arise during policy, program and project implementation due to one or combination of several reasons: (i) policies imposed without local participation, (ii) uncoordinated planning, (iii) inadequate information sharing, (iv) lack of harmony and coordination between bodies of law and legal procedures, (v) poor identification and inadequate consultation with stakeholders, (vi) lack of effective mechanisms for conflict management, (vii) limited institutional capacity and (viii) inadequate monitoring and evaluation of programs. These situations, in general, bring about conflict over resource ownership and access; conflict over decision-making associated with resource management; conflict over distribution of resource revenues as well as other benefits and burdens (UNDPA/UNEP, 2015). This especially needs a thoroughly assessment in the case of carbon projects, like REDD + program, that are more complicated. To this end, policies should be implemented in the ways that can avoid conflicts by minimizing the above mentioned deficiencies (Mery et al., 2005). The actors that involve in resource conflicts may range from local level (*e.g.*, local communities, local NGOs and local government) to international level (*e.g.*, donors and international NGOs) having either cooperative, conflictual or hybrid or changing interactions (Schilling et al., 2018).

4 Carbon Benefit Sharing

Designing a benefit-sharing mechanism without the understanding of benefits, costs, structure and distribution of incentives and benefits, beneficiaries and safeguards will fail to deliver and meet actors' interests in participating in REDD + and lead to further deforestation Angelsen et al., 2018; Chapman et al., 2014). Although benefit-sharing mechanism is at its infant stage in Ethiopia, studies show that the allocation of carbon benefits should be pro-poor with an emphasis on legal rights based on government policies or according to efforts made to reduce deforestation and forest degradation (Bekele et al., 2018; Pham et al., 2021). The seeds of carbon trading were planted during the Kyoto-Protocol in 1997 and the concept of REDD + was introduced in UNFCCC in 2008 and expanded onwards (UNFCCC 2014). Carbon trading was initiated (*e.g.*, through REDD + programs) not only to reverse tropical deforestation which accounts about 17% of global GHGs emission, but also to reverse the opinion of forests are worth more cut down than standing (IPCC, 2007).

Keeping in mind that there are variations to govern carbon rights, it is possible to create three broad categories (Foli & Dumenu, 2013; Loft et al., Peskett & Brodnig, 2011, 2015; USAID, 2012a, 2012b):

Table 1 Approaches to forest use conflict management and resolution (FAO, 2000, 2005, 2006; UNDPA/ UNEP 2015)

System	*Strategy*	*Strengths*	*Limitations*
Customary	**Negotiation** **Mediation** **Arbitration** **Coercion**	• Encourage participation by community members & respect local values and customs • Are more accessible because of their low cost, flexibility in scheduling and procedures, and use of local language • Encourage decision-making based on collaboration, with consensus emerging from wide-ranging discussions, often fostering local reconciliation • Contribute to processes of community empowerment	• Have been supplanted by courts and administrative laws • Are often inaccessible to people on the basis of gender, class, caste and other factors • Are challenged by the increasing heterogeneity of communities due to cultural change, population movements & other factors that have eroded the social relationships that supported customary conflict management • Often cannot accommodate conflicts between communities or a community and the state
National legal	**Adjudication** **Arbitration**	• Are officially established with supposedly well-defined procedures • Take national and international concerns and issues into considerations • Involve judicial and technical specialists in decision-making • Result in decisions that are legally binding	• Are often inaccessible to the poor, women, marginalized groups and to remote communities because of cost, distance, language barriers, political obstacles, illiteracy and discrimination • May not consider indigenous knowledge, local institutions and long-term community needs in decision-making • May involve judicial & technical specialists who lack the expertise, skills & orientation required for PNRM • Use procedures that are generally adversarial and promote winner-lose situation

System	*Strategy*	*Strengths*	*Limitations*
Alternative	**Negotiation Mediation**	• Promote conflict management and resolution by building upon shared interests and finding points of agreement • Involve processes which resemble those already existing in most local conflict management systems, including flexible and low-cost access • Foster a sense of ownership in the solution process of implementation • Emphasize building capacity within communities so local people become more effective facilitators, communicators, planners and handlers of conflicts.	• May encounter difficulties in getting all stakeholders to the bargaining table • May not be able to overcome power differentials among stakeholders, so that vulnerable groups such as the poor, women and indigenous people remain marginal • May result in decisions that are not legally binding • May lead some practitioners to use methods developed in other countries and settings without adapting them to local contexts

1. *Explicit carbon benefit right* is created by new law or amendment to existing law, clearly identifying the right and the right holder and how such right relates to land and forest ownership and use.
2. *Implicit carbon benefit right* is recognized based upon existing laws on land or forest rights that can be extended or interpreted to cover forest-based carbon sequestration.
3. *Contractual carbon benefit right* is based upon an agreement that stipulates the nature and scope of rights. Agreements can exist between the government and private parties, or between private parties without government involvement.
4. Combination of contractual and either of explicit or implicit carbon benefit rights.

In the Ethiopian context, the implicit approach is most commonly applied followed by the combined approach. Bale Mountains Ecoregion and Humbo Ethiopia Assisted Natural Regeneration Projects have sold, respectively, more than 150,000 carbon credits and 13,000 units of Certified Carbon Emission Reductions. Over and above, the Oromia Forested Landscape Program, with potential payments for up to 10 million tonnes of emission reductions, has received USD 18 million in grant financing (Pham et al., 2021). This shows the effects of the new REDD + policy intervention (Hildén et al., 2014; Jordan & Huitema, 2014a).

Ethiopia does not have forest law that deals with specific benefit-sharing mechanism that could be used as the starting point for conferring clear entitlements to forest carbon revenues for forest-dependent peoples except the Readiness Preparation Proposal (R-PP) reports that benefit-sharing mechanisms have been agreed between communities and governments under the backing of PFM initiatives (Hedges, 2010). Who involved in efforts to reduce deforestation and forest degradation? Who conserve and manage the forests? Can carbon fund, if it becomes real, cover all the livelihood options people deliver from forests? And how much of the revenue from carbon trading will go to the government purse and how much to the local communities? These are the leading questions to be answered regarding carbon ownership and benefit sharing (Table 2).

Carbon ownership and taking the advantages from carbon resource needs cognizant forest management that result in a quantifiable carbon sink (inc. quality data), knowledge and understanding of terms such as baseline, additionality, leakage and permanence, and a forest resource, as

Table 2 Successes and challenges experienced in Humbo community-managed carbon forestry project

Success	*challenges*
1. There was degraded community forest resource due to overexploitation because the resource was open access at which ownership is not defined 2. Negotiations have been made with local communities and government bodies to settle ownership 3. Formation of seven forest development and protection cooperatives with total target population of 48,893 4. Area mapping has been made for the whole project and then for each cooperative to avoid conflict over resource 5. Area has been rehabilitated—estimated to sequester 880,000 tCO_2e over 30 years 6. First project assessment has been made and an additional of 7,769 tCO_2e was gained from the baseline and first payment of $34,183.00 has been made by World Bank and World Vision Australia (the total budget of the project is $443,906 with additional $50,000 for capacity building). In addition to this revenue flow, soil erosion has decreased, production of forest products (fuelwood, grasses, honey) increased and these in return has improved the livelihood of the people. The most significant outcome for the community will be greater resilience to environmental shocks as well as creation of a stable income stream from forest products	1. The CDM compliance requirements made establishing the project both expensive and time consuming 2. The community struggled to understand the concept of emissions trading. Most community members had unrealistic expectations of the level of income that would be generated through carbon sales 3. The CDM registration process is long and time consuming, and the project is yet to be officially registered and recognized by the World Bank board and participating countries 4. Actual projects costs are high and cannot be afforded by the community. World Vision Australia has been investing funds until such time when carbon trading will commence 5. Ownership of the emissions reduction credits is a complicated matter since only the government is recognized as the rightful owner of the land under which the sequestration is taking place. This leaves marginal space for NGOs to participate especially under the LULUCF-a key component under CDM Initiatives which require prolonged contractual obligations. The interventions of NGOs are short-term based 6. The land tenure system and fragmentation has created setbacks with part of Humbo dropped from the partnership with the World Bank, reducing the size of the forest and the emission reduction credits significantly

(Adapted from: World Vision Ethiopia 2010)

a carbon sink, should be registered and certified for its emission reduction (Arnoldus & Bymolt, 2011; Bosa et al., 2019; FAO, 2015, 2016; Jindal & Namirembe, 2012; WVE, 2010; WB, 2016; WBG, 2016). These pre-requisites are not easy by themselves; there are other difficulties to take advantages of the carbon benefits by farmers (ICRAF, 2012) which include inability of farmers to produce the minimum volume of carbon needed for the carbon market, small and scattered landholdings on which they conduct all sorts of diverse activities to feed themselves and make a living, and that makes measuring carbon stocks a complicated business, and the costs of registering projects, drawing up contracts and monitoring stocks are exorbitantly high for smallholders acting alone or in small group.

Ethiopia's engagement in REDD + in 2008 was manifested by the submission of a Letter of Intent and REDD + Readiness Plan Program Idea Note (R-PIN) to the World Bank's Forest Carbon Partnership Facility (FCPF). Progressively, as policy diffusion (Jordan & Huitema, 2014a, 2014b), the implementation of regional REDD + programs in different regions and an emission reductions payment agreement (ERPA) for results in the Oromia program have been made effective with the help of financial contribution from donor countries (Pham et al., 2021).

Ethiopia has not passed national legislation defining tenure rights over carbon despite generally accepted the principles of benefit sharing equitably among all relevant rights holders and stakeholders with a clear stance that the REDD + safeguards take on local communities' rights and the protection of their indigenous knowledge and institutions (Bekele et al., 2015). In the context of REDD + , carbon right is the legal right to benefit from reduced greenhouse gas emissions and/or increased carbon dioxide sequestration from the forestry sector (Peskett & Brodnig, 2011; USAID, 2012b).

Payment for carbon can be considered on fairness and efficiency bases (CIFOR, 2005; MEFCC, 2018). First, carbon benefit can be paid on *fairness bases* mainly depending on negotiation; political feasibility which includes perception of fairness; legality, particularly vis-a vis land tenure and ethics, since some actors may loss illegal revenues, corrupt payoffs and unjust profit. Second, carbon benefit can also be paid on *efficiency bases:* to those who have credible site-specific claims, to communities who have a vital stake and act as direct local guardians, regional governments strengthened by decentralization regardless of the pros and cons and to those who have de facto land and resource use control capabilities (the right to exclude and can control access).

The Ethiopian R-PP has dealt important issues on how to manage forest resources and benefit from the carbon financing. However, it is deficient in explaining some main issues (Beyene et al., 2015; Duker et al., 2019; Kemerink-Seyouma et al., 2018; RDI 2010; Williams, 2013; Yami & Mekuria, 2022). These include the following:

- It is not enough simply to identify and secure the user and carbon rights of forest-dependent people but also necessary to have a mechanism by which they can receive their equitable share of carbon benefits.
- The REDD + Social and Environmental Standards insist that REDD + benefits be shared "equitably" by all stakeholders, but the R-PP provides for such benefits to be shared "equally" which is, of course, very different standards.
- The R-PP emphasizes the stakeholder consultation and participation as key issue, that is true, but the right of stakeholders to full information and to say no if things are uncomfortable has not incorporated into the plan.
- Often customary tenure rights are essentially community tenure rights, but it is largely missing from the R-PP.
- The R-PP recognizes the need to ensure the long-term livelihood security and well-being of local forest-dependent communities but pays little attention to vulnerable groups.
- The R-PP deals to some extent that women are able to participate meaningfully in the stakeholder consultations, but did not deal how women's tenure rights are identified and protected.

5 Ethiopian Government Efforts to Address the Carbon Ownership Challenges

The government of Ethiopia has given pronounced attention for the forestry sector as it plays a central role in realizing the country's commitment to achieving the country's Climate-Resilient Green Economy (CRGE). Given the recognition that land use is the source of 88% GHG emissions, a core element of the CRGE's vision is sustainable landscape management and interventions in the land use sector contribute to social and environmental objectives, such as food security, watershed and biodiversity protection. The government has developed forest sectors development program dealing on sustainable forest production and value chains, forest environmental services, and forests and rural livelihoods which expected to generate 3.3 dollars of benefit for every dollar invested (MEFCC, 2018). The government has made policy reforms in the forest sector. For instance, in Forest Development, Conservation and Utilization Proclamation (Proclamation No. 1065/2018) had set four forest ownerships types (state, private, community and association forests). Community and association forests were excluded in Forest Development, Conservation and Utilization Proclamation (Proclamation No. 542/2007). In the recent policy Proclamation (No. 1065/2018) article two, sub-articles 7 and 8, define community forest as a forest developed, conserved, utilized and administered by the community on its private or communal possession based on bylaws and plans developed by the community, and association forest as a forest developed, conserved, utilized and administered by associations established to develop forest. This policy reform hopefully resolves the carbon ownership challenges in Ethiopia.

There is also improvement of community ownership on reforestation and afforestation sites which are being supported by local bylaws. Forest access right of the community is legally secured through PFM (Ali & Bachano, 2020). PFM has been introduced in a landscape context strategy to avoid leakage making it one of the solutions to effect REDD + strategies (Beyene et al., 2015; FAO, 2010b; Lemenih et al., 2015). PFM is also one of the solutions to define property relations to the forest resources, including carbon, and builds forest management and conservation as well as power sharing thereby enhancing carbon benefit in the REDD + strategies (Ameha et al., 2014; Ayana et al., 2015; Gobeze et al., 2009; Jatana & Paulos, 2017).

6 Conclusions

The essence of this paper is to highlight some policy direction to be taken in the area of forest and carbon governance in Ethiopia that helps the effective implementation of REDD + . Some of the policy directions should include:

- The decentralization process in Ethiopia must overcome corruption and establish new structures of governance at the local level.
- Mechanisms should be created to incorporate traditional knowledge into policy-making processes (close gaps between *de jure* and de factorights).

- Careful monitoring and reporting must take place on the national forest program's processes to insure that the outcome of these processes accomplishes substantive and desirable policy change and is not purely symbolic in order to certainly be a core components of any international governance regime.
- Natural resource management, including forests and carbon-related policies, should be based on sound scientific evidence and reliable information, and research and development funding must be maintained.
- National-based expert panel on forests, similar in nature to the IPCC should be established so that it could support international forest policy deliberations, provide a quick response to emerging regional or international forest-related issues and bring them to global attention.
- Government must have carbon policy amalgamated with forest policy and be committed to its implementation.
- Natural resource policies should be reviewed to enable involvements of communities and other stakeholders in management and utilization of forest resources including carbon.
- There is a need for forest policy reviews so that threatened forests as open access should be rebuilt.
- Land policies should be reviewed so as to enable communities and villagers to have secure and clear tenure rights.
- Policies that have impacts on forests should be reviewed to address issues of incompatibility.
- Tenure should be a start-point than an afterthought in carbon ownership arrangements such as REDD + .
- There is a need to close the gap between rights on paper and rights in practice.

References

ADF (African Development Forum) 2010. Governance and leadership responses to climate change. *Seventh African Development Forum: Acting on Climate Change for Sustainable Development in Africa*. Issue Paper No. 1. Addis Ababa, Ethiopia.

Alemie, B. K., & Amsalu, T. (2020). Does land tenure insecurity affect forest cover change? evidence from Gerejeda State forest in Ethiopia. *Journal of Land and Rural Studies, 8*(2), 101–120.

Ali, A., & Bachano, T. (2020). Review on the role of participatory forest management on livelihoods of local community in Ethiopia. *Journal of Forestry and Environment, 2*(2), 36–45.

Ameha, A., Larsen, H. O., & Lemenih, M. (2014). Participatory forest management in Ethiopia: Learning from pilot projects. *Environmental Management, 53*, 838–854. https://doi.org/10.1007/s00267-014-0243-9

Andoh, J., & Lee, Y. (2018). National REDD+ Strategy for climate change mitigation: A review and comparison of developing countries. *Sustainability, 10*(4781). https://doi.org/10.3390/su10124781.

Angelsen, A. (2008). (Ed.). *Moving ahead with REDD: Issues, options and implications*. CIFOR, Bogor.

Arnoldus, M. W., & Bymolt, R. (2011). *Demystifying carbon markets: A guide to developing carbon credit projects*. KIT Publishers.

Ayana, A. N., Arts, B., & Wiersum, K. F. (2013). Historical development of forest policy in Ethiopia: Trends of institutionalization and deinstitutionalization. *Land Use Policy, 32*, 186–196.

Ayana, A. N., Vandenabeele, N., & Arts, B. (2015). Performance of participatory forest management in Ethiopia: Institutional arrangement versus local practices. *Critical Policy Studies*. https://doi.org/10.1080/19460171.2015.1024703

Bekele, M. (2003). *Forest property rights, the role of the state and institutional exigency: The Ethiopian experience*. PhD Thesis, Swedish University of Agricultural Sciences, Uppsala, Sweden.

Bekele, M., Tesfaye, Y., Mohammed, Z., Zewdie, S., Tebikew, Y., Brockhaus, M., & Kassa, H. (2015). *The context of REDD+ in Ethiopia: Drivers, agents and institutions.* Occasional Paper 127. CIFOR. Bogor, Indonesia.

Bekele, M., Zewdie, S., Boissière, M., & Atmadja, S. (2018). *REDD+ MRV implementation in Ethiopia: Review of the context, framework and progress.* Occasional Paper 192. Bogor, Indonesia: CIFOR.

Beyene, A. D., Bluffstone, R., & Mekonnen, A. (2015). Community forests, carbon sequestration and REDD+: Evidence from Ethiopia. *Environment and Development Economics, 21*(2), 249–272.

Birhan, E., Assefa, E., & Petrova, M. A. (2021). Challenges of forest governance in addressing REDD+: Status, effects and prospects. The case of Bale Eco-Region, Oromia Regional State, Ethiopia. *Geography, Environment, Sustainability 14*(1): 185–195.

Bosa, A. B., De Sya, V., Duchelleb, A. E., Herolda, M., Martiusb, C., & Tsendbazar, N. E. (2019). Global data and tools for local forest cover

loss and REDD+ performance assessment: Accuracy, uncertainty, complementarity and impact. *International Journal of Applied Earth Observatory and Geoinformation, 80*, 295–311.
C2ES (Centre for Climate and Energy Solutions). (2019). *Climate Essentials: What is climate resilience and why does it matter?* Arlington.
Chapman, S., Wilder, M., & Millar, I. (2014). Defining the legal elements of benefit sharing in the context of REDD+. *Carbon and Climate Law Review, 8*(4), 270–281.
CIFOR (Center for International Forestry Research). (2000). *The underlying causes of forest decline*. Occasional Paper No. 30. Jakarta, Indonesia.
CIFOR. (2005). *Payments for Environmental Services: Some nuts and bolts*. CIFOR Occasional Paper No. 42. Jakarta, Indonesia.
CIFOR. (2008). *Monitoring Forest Emissions: A Review of Methods*. Working Paper No. 39. Bogor, Indonesia.
CIFOR. (2009). *Recognizing community rights: The potential and challenges of forest tenure reform*. Info Brief No. 22, Indonesia.
CIFOR. (2010). *REDD, forest governance and rural livelihoods: The emerging agenda*. Bogor, Indonesia.
Corbera, E., Estrada, M., May, P., Navarro, G., & Pacheco, P. (2011). Rights to land, forests and carbon in REDD+: Insights from Mexico, Brazil and Costa Rica. *Forests, 2*, 301–342.
Cotula, L., & Mayers, J. (2009). *Tenure in REDD—Start-point or afterthought?* Natural Resource Issues No. 15. International Institute for Environment and Development. London, UK.
Deacon, R. T., & Mueller, B. (2006). Political Economy and Natural Resource Use. In R. López & M. A. Toman (Eds.), *Economic development and environmental sustainability: New policy options* (pp. 122–153). Oxford University Press Inc.
Dilling, L. (2007). Toward Carbon Governance: Challenges across scales in the United States. *Global environmental politics, 7*(2), 28–44.
Duker, A. E. C., Tadesse, T. M., Soentoro, T., de Fraiture, C., & Kemerink-Seyoum, J. S. (2019). The implications of ignoring smallholder agriculture in climate-financed forestry projects: Empirical evidence from two REDD+ pilot projects. *Climate Policy, 19*(s1), S36–S46.
FAO. (2000). *Conflict and natural resource management*. Italy.
FAO. (2005). *Negotiation and mediation techniques for natural resource management*. Italy.
FAO. (2006). *Land tenure alternative conflict management*. Land tenure manuals 2, FAo, Rome, Italy.
FAO. (2010b). *Participatory forest management in Ethiopia: Practices and experiences*. FAO Sub Regional Office for Eastern Africa (SFE), Addis Ababa, Ethiopia.

FAO. (2010a). *Global forest resources assessment 2010a: Country Report, Ethiopia.* FAO, Rome, Italy.

FAO. (2015). *Technical considerations for Forest Reference Emission Level and/or Forest Reference Level construction for REDD+ under the UNFCCC.* Available at (01/08/2022):https://unfccc.int/files/land_use_and_climate_change/application/pdf/redd_20150804_unredd_technical_considerations_frel_under_unfccc_en.pdf

FAO. (2016). *Forestry for a low-carbon future: Integrating forests and wood products in climate change strategies.* FAO FORESTRY PAPER 177, Rome, Italy.

FAO. (2020). *Global Forest Resources Assessment 2020, Ethiopia Country Report*, Rome, Italy.

FDRE (Federal Democratic Republic of Ethiopia). (1994). *The Constitution of The Federal Democratic Republic of Ethiopia, 1994.* Addis Ababa, Ethiopia.

FDRE. (1997). *The Federal Rural Land Proclamation 89/1997.* Addis Ababa, Ethiopia.

FDRE. (2007). Forest Development, Conservation and Utilization Proclamation. *Negarit Gazeta,* proclamation # 542/2007. Addis Ababa, Ethiopia.

FDRE. (2018). Forest Development, Conservation and Utilization Proclamation. *Negarit Gazeta,* proclamation # 1065/2018. Addis Ababa, Ethiopia.

Foli, E. G., & Dumenu, W. K. (2013). *Background Paper for Ghana's REDD+ Benefit Sharing Dialogue.* CSIR-Forestry Research Institute of Ghana, *Elmina, Ghana.*

Gibbs, H. K., & Brown, S. (2007a). *Geographical distribution of woody biomass carbon stocks in Tropical Africa: An updated database for.* https://doi.org/10.3334/CDIAC/lue.ndp055.2007

Gizachew, B., Astrup, R., Vedeld, P., Zahabu, E. M., & Duguma, L. A. (2017). REDD+ in Africa: Contexts and challenges. *Natural Resources Forum.* https://doi.org/10.1111/1477-8947.12119

Gobeze, T., Bekele, M., Lemenih, M., & Kassa, H. (2009). Participatory forest management and its impacts on livelihoods and forest status: The case of Bonga forest in Ethiopia. *International Forestry Review, 11*(3), 346–358.

Hairiah, K., Dewi, S., Agus, F., Velarde, S., Ekadinata, A., Rahayu, S., & van Noordwijk, M. (2010). *Measuring carbon stocks across land use systems: A Manual.* Bogor, Indonesia. World Agroforestry Centre (ICRAF), SEA Regional Office, pp155.

Hanna, S., Folke, C., & Mäler, K. G. (Eds.). (1996). Property rights and the natural environment. In *Rights to nature: Ecological, economic, cultural and political principles of institutions for the environment*, Island Press.

Hardin, G. (1968). The tragedy of the commons. *Science, New Series, 162*(3859), 1243–1248.

Hedges, A. (2010). *Forest carbon rights in REDD+ countries: A snapshot of Africa*. Norton Rose LLP, London, UK.

Hildén, M., Jordan, A., & Rayner, T. (2014). Climate policy innovation: Developing an evaluation perspective. *Environmental Politics, 23*(5), 884–905.

HLPE. (2011). *Land tenure and international investments in agriculture*. A report by the High Level Panel of Experts on Food Security and Nutrition of the Committee on World Food Security, Rome, Italy.

Houghton, R. A. (1999). The annual net flux of carbon to the atmosphere from changes in land use 1850–1990. *Tellus B, 51*, 298–313.

ICRAF. (2012). *Trees, livelihoods, landscapes*. Annual Report 2011–2012, Nairobi, Kenya.

IPCC. (2007). *IPCC Fourth Assessment Report: Climate Change 2007*. Switzerland.

IPCC. (2022). *Summary for policymakers. Climate change 2022: Impacts, adaptation, and vulnerability*. Contribution of Working Group II to the Sixth Assessment Report of the Intergovernmental Panel on Climate Change. Cambridge University Press (In Press).

Jatana, G., & Paulos, Z. (2017). Farmer's participation in participatory forest management and factors affecting its performance: The case of Sodo Zuriya District, Wolaita Zone, Ethiopia. *Journal of Economics and Sustainable Development, 8*(9), 52–62.

Jordan, A., & Huitema, D. (2014a). Innovations in climate policy: Conclusions and new directions. *Environmental Politics, 23*(5), 906–925.

Jordan, A., & Huitema, D. (2014b). Policy innovation in a changing climate: Sources, patterns and effects. *Global Environmental Change, 29*, 387–394.

Kemerink-Seyouma, J. S., Tadessea, T. M., Mershaa, W. K., Dukera, A. E. C., & Fraiture, C. D. (2018). Sharing benefits or fueling conflicts? The elusive quest for organizational blue-prints in climate financed forestry projects in Ethiopia. *Global Environmental Change, 53*, 265–272.

Larson, A. M. (2012). *Tenure rights and access to forests: A training manual for research*. CIFOR, Bogor, Indonesia.

Larson, A. M., & Petkova, E. (2011). An Introduction to Forest Governance, People and REDD+ in Latin America: Obstacles and Opportunities. *Forests, 2*, 86–111.

Lemenih, M., Allan, C., & Biot, Y. (2015). *Making forest conservation benefit local communities: Participatory forest management in Ethiopia*. FARM Africa, London, UK.

Loft, L., Pham, T. T., & wong GY, Brockhaus M, Le DN, Tjajadi JS, Luttrell C,. (2017). Risks to REDD+: Potential pitfalls for policy design and implementation. *Environmental Conservation, 44*(1), 44–55.

Loft, L., Ravikumar, A., Gebara, F. M., Pham, T. T., Resosudarmo, I. A. P., Assembe, S., Tovar, J. G., Mwangi, E., & Andersson, K. (2015). Taking stock of carbon rights in REDD+ Candidate Countries: Concept meets reality. *Forests, 6*, 1031–1060.

MEFCC. (2018). *National REDD+ Secretariat, National REDD+ Strategy 2018–2030*. Final Report. Addis Ababa, Ethiopia.

MEFCC (Ministry of Environment, Forest and Climate Change). (2018). National Forest Sector Development Program, Ethiopia Volume II: Program Pillars, Action Areas and Targets. Addis Ababa, Ethiopia.

Meinzen-Dick, R., & Mwangi, E. (2008). Cutting the web of interests: Pitfalls of formalizing property rights. *Land Use Policy, 26*, 36–43.

Mery, G., Alfaro, R., kanninen, M., Lobovikov, M., & Pye-Smith, C. (eds.) (2005). *Forests for the new millennium: Making forests work for people and nature*. Ministry for Foreign Affairs of Finland and International Union of Forest Research Organizations, Helsinki, Finland.

Naughton-Treves, L., & Wendland, K. (2013). Land tenure and tropical forest carbon management. *World Development*. https://doi.org/10.1016/j.worlddev.2013.01.010

Nune, S., Kassie, M., & Mungatana, E. (2009). *Forestry resource accounting: The experience of Ethiopia*. In: Proceedings of the National Workshop on Sustainable Land Management and Poverty Alleviation. Addis Ababa, Ethiopia.

Peskett, L., & Brodnig, G. (2011). *Carbon rights in REDD+: Exploring the implications for poor and vulnerable people*. World Bank and REDD-net. Washington DC, USA.

Pham, T. T., Moeliono, M., Dwisatrio, B., Yuwono, J., & Atmadja, S. (2021). REDD+ benefit sharing in Ethiopia: Policy and stakeholder perceptions analysis. *International Forestry Review, 23*(4), 476–491.

RDI (Rural Development Institute). (2010). *Review of the Ethiopia's draft Readiness Preparation Proposal (R-PP) for Reducing Emissions from Deforestation and Degradation (REDD); implications for land, forest and carbon tenure*. Seattle.

Rijal, S., Subedi, M., Chhetri, R., & Joshi, R. (2021). Good governance assessment in community forest of Nepal. *Journal of Forest and Environmental Science, 37*(3), 251–259.

Robinson, B. E., Holland, M. B., & Naughton-Treves, L. (2011). *Does secure land tenure save forests? A review of the relationship between land tenure and tropical deforestation*. CCAFS Working Paper no. 7. CGIAR Research Program on Climate Change, Agriculture and Food Security (CCAFS). Copenhagen, Denmark.

RPP (The Readiness Preparation Proposal–Ethiopia). (2010). *Draft report*, Addis Ababa, Ethiopia.

Schilling, J., Saulich, C., & Engwicht, N. (2018). A local to global perspective on resource governance and conflict. *Conflict, Security & Development, 18*(6), 433–461.

Sunderlin, W., Larson, A. M., Cronkleton, P. (2009). Forest tenure rights and REDD+: From inertia to policy solutions, pp. 139–150. In A Angelsen, (Ed.), *Realizing REDD+: National strategy and policy options*, CIFOR: Bogor.

Tompkins, E. L., Nicholson-Cole, S. A., Hurlston, L. A., Boyd, E., Hodge, G. B., Clarke, J., Gray, G., Trotz, N., Varlack, L. (2005). *Surviving Climate Change in Small Islands – A Guidebook*, Tyndall center for Climate Change Research, UK.

UNCCD (United Nations Convention to Combat Desertification). (2017). *Land tenure and rights for improved land management and sustainable development*. Global Land Outlook Working Paper.

UNDPA/ UNEP (United Nations Department of Political Affairs and United Nations Environment Programme). (2015). *Natural Resources and Conflict: A Guide for Mediation Practitioners*. New York, USA and Nairobi. Kenya.

USAID. (2012a). *Forest carbon rights guidebook: A tool for framing legal rights to carbon benefits generated through REDD+ programming*, Washington DC, USA.

USAID. (2012b). *REDD+ and carbon rights: lessons from the field property rights and resource governance project (PRRGP)*, Washington DC, USA.

Vreugdenhil, D., Payton, I. J., Vreugdenhil, A., Tilahun, T., Nune, S., & Weeks, E. (2011). *Establishing a "Carbon Baseline" and Mechanisms for Payments for Carbon Environmental Services Discharged by Protected Areas in Ethiopia*. World Institute for Conservation and Environment.

WB (World Bank). (2016). "Carbon Credits and Additionality: Past, Present, and Future." PMR Technical Note 13. Partnership for Market Readiness, World Bank, Washington, DC. USA.

WBG (World Bank Group). (2016). Networked Carbon Markets: Networking Carbon Markets— Key Elements of the Process.

WBISPP. (2005). *A national strategy plan for the biomass sector*. Addis Ababa, Ethiopia.

Williams, L. G. (2013). *Putting the pieces together for good governance of REDD+: An analysis of 32 REDD+ Country Readiness Proposals*. Working Paper. World Resources Institute, Washington, DC, USA.

Wodajo, A., Mohammed, M., & Tesfaye, M. A. (2020). Carbon stock variation along altitudinal and slope gradients in Gara-Muktar forest, West Hararghe zone, Eastern Ethiopia. *Forestry Research and Engineering: International Journal, 4*(1), 17–26.

WRI (World Resources Institute). (2002). *Closing the Gap: Access to Information, Participation, and Justice in Decision-Making for the Environment*. Washington, DC.USA.

Yami, M., & Mekuria, W. (2022). Challenges in the governance of community-managed forests in Ethiopia: Review. *Sustainability, 14*, 1478. https://doi.org/10.3390/su14031478

CHAPTER 10

Climate Variability and Household Food Security in the Guinea Savannah Agro-Ecological Zone of Ghana

Seth Opoku Mensah, Timothy Amang-bey Akanpabadai, Michael Addaney, Seth Asare Okyere, and Stephen Kofi Diko

1 Introduction

After significant achievements in reducing global hunger, alarm bells have been raised as estimates of world hunger reversed course and began to increase in 2015, for the first time in more than 15 years (Food and Agriculture Organization [FAO] et al., 2017). The world is, therefore, not on track to eradicate hunger by 2030, as evidence continues to point to a rise in world hunger in recent years (FAO et al., 2018; FAO et al., 2021). There are indications that the main drivers behind this observed

S. Opoku Mensah (✉)
Institute for Sustainable Futures, University of Technology Sydney, Sydney, NSW, Australia
e-mail: Seth.Opokumensah@student.uts.edu.au

T. A. Akanpabadai
Catholic Relief Services, Tamale, Northern Region, Ghana
e-mail: timothy.amang-beyakanpabadai@crs.org

M. Addaney et al. (eds.), *Climate Change in Africa*,
https://doi.org/10.1007/978-3-031-30050-9_10

reversal of world hunger are related to the changing climate. This has brought increased urgency to question the role of climate variability and extremes such as heat spells, floods, droughts, and storms (Holleman et al., 2020). In this chapter, climate variability is the variations in the mean state and other statistics (standard deviations, extremities, etc.) of the climate on all spatial and temporal scales beyond that of individual weather events. It may be due to natural internal processes within the climate system (internal variability) or natural or anthropogenic external forcing (external variability) variations. Climate extreme is the occurrence of a value of a weather or climate variable above or below a threshold value near the upper or lower ends of the range of observed values of the variable. For simplicity, extreme weather and climate events are referred to collectively as climate extremes (Intergovernmental Panel on Climate Change [IPCC], 2012).

Increasing climate variability and extremes create multiple and compounding impacts on food systems and are already negatively impacting food security; a situation that exists when all people, at all times, have physical, social, and economic access to sufficient, safe, and nutritious food that meets their dietary needs and food preferences for an active and healthy life. This is making the challenge of ending hunger and malnutrition even more difficult (Holleman et al., 2020; FAO et al., 2018) due to the exposed and sensitive nature of food production systems from increasing temperatures, precipitation change, and intensified extreme events in the face of limited capacity to adapt (IPCC, 2022; United Nations Framework Convention on Climate Change [UNFCCC],

M. Addaney
Department of Land Management, University of Energy and Natural Resources, Sunyani, Ghana
e-mail: michael.addaney@uenr.edu.gh

S. A. Okyere
College of Architecture, Planning and Landscape Architecture, University of Arizona, Tucson, AZ, USA
e-mail: sethaokyere@arizona.edu

S. K. Diko
University of Memphis, Memphis, TN, USA
e-mail: skdiko@memphis.edu

2021; World Bank, 2021). As detailed in Table 1, increasing climate variability and extremes affect all four dimensions of food security (FAO et al., 2021; Weldearegay & Tedla, 2018; Holleman et al., 2020) and act as multiplier of existing threats to food security (Alemu & Mengistu, 2019; Abegunde et al., 2022; Ahn et al., 2022).

Table 1 Climate implications on food security

Food security dimension	*Description*	*Climate implications*
Food availability	Food availability at the household level can be described as its physical existence, either from markets or own production, including food imports and aid	Climate variability and extremes affect food availability through changes in production. Extreme weather events disrupt production cycles
Food accessibility	It is a household's capacity to access food of the needed quantity, quality, and nutritional type through production, purchase, gifts, borrowing, or aid. It is directly associated with a household's financial ability to afford the required amount of food for an active and healthy life	Limited food production in the face of climate variability and extremes leads to food scarcity, high prices, and limits accessibility. Increased exposure to heavy floods and prolonged droughts brought about by climate variability and extremes reduces households' entitlement and hinders food access
Food utilization	Utilization shows whether individuals and households use well the food they have access to and whether they consume well-balanced and nutritious foods they can easily afford or choose a nutritionally low diet. It also shows whether the foods they consume are safe and well-cooked, giving their total nutritional value and substantive nutrients	Climate variability and extremes affect the rate of production of different food types. This may hinder the dietary requirement of households. It can also affect the income levels and the ability of households to attain the required foods to meet a nutritious and well-balanced diet. When food is unavailable, the variety of food items typically decreases. This can reduce the number of meals and hence a low balance of nutrients

(continued)

Table 1 (continued)

Food security dimension	*Description*	*Climate implications*
Food stability	If food availability, access, and utilization are sufficiently met, stability is the condition in which the whole system is stable, thus ensuring that households are food secure at all times	Climatic variability and extremes can upset the stability of households' food security strategies and create fluctuations in food availability, access, and utilization. Climatic variability and extremes can lead to unstable harvests, instability of food production, and destabilize local food systems

Source FAO, 2015; Baldos & Hertel, 2014; FAO et al., 2014; United States Agency for International Development (USAID), 2013; Belloumi, 2014; Poppy et al., 2014; Weldearegay & Tedla, 2018; FAO et al., 2021

2 Climate Variability and Extremes, and Food Security in the Developing World

Climate variability and extremes are the main drivers of food insecurity in the developing world. They affect the productivity of the agriculture sector, its stability, and other components of the food system, including storage, access, and utilization (Ali et al., 2017; Ayinu et al., 2022). The climate implications are significantly worse in countries with agri-food systems that are highly exposed and sensitive to rainfall and temperature variability and extremes including where a high proportion of the population depends on agriculture for livelihood (Holleman et al., 2020; FAO et al., 2018) such as exist in developing countries (Weldearegay and Tedla, 2018). The number of developing countries exposed to climate variability and extremes has steadily increased over the past 20 years, from 76 percent in 2000–2004 to 98 percent in 2015–2020 (FAO et al., 2021). The vulnerabilities of developing countries are caused by low level of technological progress, lack of resources to mitigate the adverse impacts of climate variability and extremes on agriculture, and their greater dependence on climate-sensitive natural resources and traditional agricultural practices for subsistence and livelihoods (United Nations Development Programme [UNDP], 2018; United Nations Environment Programme [UNEP], 2021; IPCC, 2018).

Due to the frequency and intensity of extreme events such as droughts, floods, storms, and climate-induced pests and plant diseases (IPCC, 2019; Phiiri et al., 2016), food security continues to remain a far-fetched dream for many people, particularly in sub-Saharan Africa, where they are highly dependent on rain-fed agriculture (Adane et al., 2015; Phiiri et al., 2016). The changing climate threatens food security in several ways. For instance, changes in rainfall leading to drought or flooding directly affect crop production and/or warmer or cooler temperatures leading to changes in the length of the growing season, and changes in markets, food prices and supply chain infrastructure (IPCC, 2014). Additionally, climate variability and extremes threaten the production and distribution of food, endanger food accessibility by destabilizing food prices and harming people's health by affecting their nutrition and diets and the quality and quantity of food produced (FAO, 2013). There is, therefore, no better way of adapting to climate change tomorrow than adapting to climate variability and extremes today. A useful way to prepare for uncertain future climate conditions is learning from current climate variability and extremes and implementing suitable policy interventions to improve resilience of farmer households (Arribas et al., 2011).

Among the agricultural stakeholders who are disproportionally affected by climate variability and extremes are smallholder farmers who lead subsistence lives and constitute over 80% of the farming population in the developing world (Awazi et al., 2022; FAO et al., 2018). Ensuring food security amidst climate variability and extremes is, therefore, a priority in developing countries such as Ghana where food insecurity is real (Asare-Nuamah, 2021). In Ghana, climate and food literature is replete with food production (e.g., Firdaus et al., 2019), general adaptation approaches in the agriculture sector (e.g., Tangonyire & Akuriba, 2021; Antwi-Agyei et al., 2014), adaptation practices of smallholder farmers (e.g., Aniah et al., 2019; Kumasi et al., 2019) and farmers' perceptions, and climate change adaptation (e.g., Callo-Concha, 2018). However, the impacts of climate variability and extremes on food security in Ghana, especially from access perspective, have received little to no attention (Asare-Nuamah, 2021). This is particularly worrying within the study setting which has the highest proportion of food insecure households in the Upper East region of Ghana (World Food Programme [WFP], Ghana Statistical Service [GSS], and Ministry of Food and Agriculture [MoFA], 2012). The chapter, therefore, contributes to narrowing this research gap by applying the Household Food Insecurity Access Scale

(HFIAS) as a proxy of food security (access) among crop-livestock smallholder farmer households in the Talensi district of the semi-arid guinea savannah agro-ecological zone of Ghana.

3 Agro-Ecological Context of the Study Area: Talensi District

Based on initial engagements with the Upper East Region's Department of Agriculture (DoA) and the Ghana Meteorological Agency (GMA), it was found that the Talensi district (see Fig. 1) is one of the most exposed and sensitive districts in Ghana to climate variability and extremes with limited adaptive capacity. It was, therefore, selected for the research. The district has a unimodal rainfall pattern with predominance rain-fed agri-food systems by smallholder farmers. It is a drought- and famine-prone, and environmentally fragile dryland (non-irrigated) area (World Vision Australia [WVA], 2018; Kandel et al., 2021), where people mainly derive their livelihoods from subsistence agriculture characterized by crop-livestock production systems on fragmented, degraded, and overutilized lands. It has a total land area of 838.4 km^2 with 81,194 people (50.3% males and 49.7% females). The district has a tropical climate with two distinct seasons: an erratic rainy season (May–October) and a dry season (November–April). It has a mean annual rainfall of 95 mm, a maximum temperature of 45 °C in March and April, and a minimum of 12 °C in December (GSS, 2014) (Fig. 1).

Agriculture employs about 78.7 percent of the district's labor force, who are primarily involved in the cultivation of sorghum, millet, maize, guinea-corn, beans, groundnut, onions, tomatoes, and rearing of goat, sheep, cattle, and poultry, among others (GSS, 2014; Kandel et al., 2021). Industrial activities are generally low in the district (Tangonyire & Akuriba, 2021) and include processing shea and *Parkia biglobosa*/African locust bean (GSS, 2014). According to WFP, GSS, and MoFA (2012), the district has the highest proportion of food insecure households (39.1%) in the Upper East region. It has much higher levels of poverty (based on wealth index scores) and very high reliance on cash purchases for food (GSS, 2018; MoFA, 2019). In recent decades, its human population of crop-livestock farmers have seen annual rainfall decline, forest cover and associated biodiversity disappear, and loss of productivity from declining soil fertility (Kandel et al., 2021; WVA, 2018).

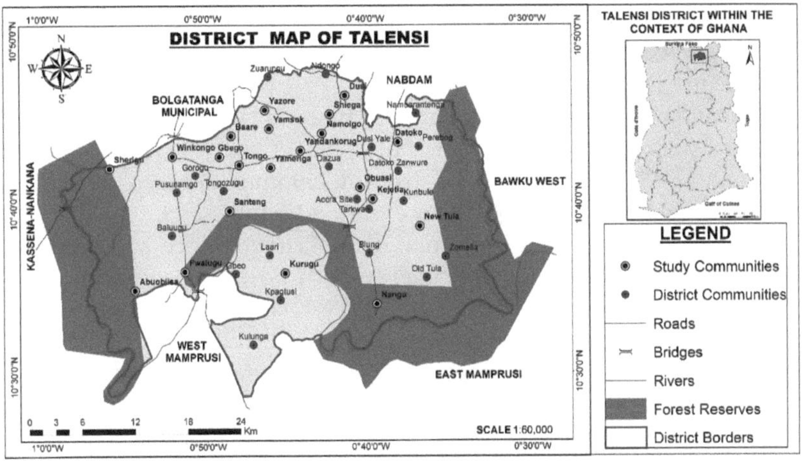

Fig. 1 The Talensi district map showing study communities (*Source* Authors' Construct, 2022)

4 Research Approach

4.1 Data Sources and Collection Procedures

The study employed a cross-sectional research design with a mixed-method approach that spanned eight months (May to December 2020) of fieldwork. The timing of the study from the start of the rainy season into the lean season offered the researchers the opportunity to appreciate how the crop-livestock farming households coped with food insecurity, especially after the farming season (see Ahn et al., 2022). The mixed-method approach was used because climate variability and extremes and food security cannot be solely studied either quantitatively or qualitatively. A blend of both approaches enhances a deeper understanding of the subject matter, enriches the study (Creswell, 2014), and provides an avenue to triangulate the research findings. For most of the fieldwork, the researchers lived and worked with the smallholder farmers and engaged in daily farming practices with them. This allowed the collection of quantitative and qualitative data from the natural setting of research participants (Asare-Nuamah, 2021). Data was collected through household surveys, focus group discussions (FGDs), key informant interviews (KIIs), and observations. This allowed the study participants to share their lived experiences, practices, and knowledge on climate variability and extremes and food security.

The study was implemented in three phases. The first phase involved selecting study communities. With the help of staff from the district's DoA and GMA, 21 crop-livestock farming communities (see Fig. 1) that are highly sensitive and exposed to climate variability and extremes with less capacity to adapt were selected. To collect adequate and representative data and provide valuable context for in-depth analysis, seven (7) communities were randomly selected from each of the district's three area councils. Reconnaissance visits were then conducted to verify the accuracy of their selection, familiarize and establish necessary contacts, and acquire entry into each of them. Community and farm transect walks were undertaken to observe the communities' settings and farming practices.

The second phase involved identifying the unit of analysis and sample size determination for the household survey. Crop-livestock smallholder farmer households were naturally identified as the unit of analysis using the. To sample the population with the desired degree of accuracy (Stephanie, 2003), the Slovin's formula (Slovin, 1960):

$$n = \frac{N}{1 + N(\alpha)^2}$$

where n = sample size, N = sampling frame (total household size [16,186]), and α = margin of error (0.05) were used to select 390 crop-livestock farmer households. The crop-livestock farmer household heads were randomly selected to obtain the required sample sizes for each of the 21 crop-livestock farming communities. Because the total household sizes were inconsistent across the 21 study communities, the sample of each community was selected based on their household sizes. Also, in the second phase, questionnaires were administered, which focused primarily on household farming practices, crop yields, food supply, and consumption, perceived impact of climate variability and extremes on household food security. Data was collected using a questionnaire coded with Open Data Kit (ODK) software. The questionnaire was pre-tested and modified where necessary. Each household survey took an average of 30 to 45 minutes to administer.

Based on the initial analysis of the household data, FGDs and KIIs were conducted in the third phase of data collection. Three FGDs were conducted per community and comprised 5 to 10 discussants. Each focus group comprised male-only, female-only, and youth-only groups. This was informed by the patriarchal nature of the study communities that has resulted in power dynamics between men, women, and youth which prevent women and youth from expressing themselves and speaking up in mixed groups to capture differences in perspective. The discussions took an average of three hours and centered on household farming practices, crop yields, food supply, and consumption, perceived impact of climate variability and extremes on household food security. The research team moderated the focus groups with the support of local research assistants who acted as interpreters and note takers/recorders. The discussions were conducted in the local dialects of the communities: Talen and Guruni. All discussions were digitally recorded with the consent of the discussants. This was followed by KIIs. A total of 63 participants (3 per community) were selected as key informants for in-depth interviews on issues raised during the FGDs. The key informants were farmers with in-depth knowledge of the subject matter and over 20 years of farming experience.

The study adopted HFIAS as a proxy of food security (access). HFIAS consists of two types of questions: nine occurrence and nine frequency-of-occurrence questions (See Table 2). The respondents were first asked if a given condition was experienced (yes or no) within 30-day recall period, and if it was, with what frequency (rarely, sometimes, or often) (Coates et al, 2007; INDDEX Project, 2018). Each household's responses place them in a single, unique category, as shown in the categorization scheme in Table 2 (Coates et al., 2007). HFIAS provides a continuous measure of household food insecurity which can be categorized into four levels of household food insecurity (food secure, mildly food insecure, moderately food insecure, and severely food insecure). According to Coates, Swindale, and Bilinsky (2007, p. 18) and as illustrated in Table 2, 'Households are categorized as increasingly food insecure as they respond affirmatively to more severe conditions and/or experience those conditions more frequently. A food secure household experiences none of the food insecurity (access) conditions or just experiences worry, but rarely. A mildly food insecure (access) household worries about not having enough food sometimes or often, and/or is unable to eat preferred foods, and/or eats a more monotonous diet than desired and/or some foods considered undesirable, but only rarely. But it does not cut back on quantity nor experience any of three most severe conditions (running out of food, going to bed hungry, or going a whole day and night without eating).

A moderately food insecure household sacrifices quality more frequently by eating a monotonous diet or undesirable foods sometimes or often, and/or has started to cut back on quantity by reducing the size of meals or number of meals, rarely or sometimes. But it does not experience any of the three most severe conditions. A severely food insecure household has graduated to cutting back on meal size or a number of meals often and/or experiences any of the three most severe conditions (running out of food, going to bed hungry, or going a whole day and night without eating), even as infrequently as rarely. In other words, any household that experiences one of these three conditions even once in the last four weeks (30 days) is considered severely food insecure.'

4.2 Data Analysis

Following Coates, Swindale, and Bilinsky (2007), the HFIAS score variable was calculated for each household. This was done by summing the codes for each frequency-of-occurrence question, as illustrated in Eq. 1. The maximum score for a household is 27 where a household response to all nine frequency-of-occurrence questions was 'often' and coded 3; the minimum score is 0 where a household responded 'no' to all occurrence questions; and the frequency-of-occurrence was subsequently coded as 0. The higher the score, the more food insecurity (access) the household experienced. The lower the score, the less food insecurity (access) a household experienced.

HFIAS Score

$$\begin{aligned} & Sum\ frequency - of - occurrence\ question\ response\ code \\ & = \text{sum (Q1a + Q2a + Q3a + Q4a} \\ & \quad \text{+ Q5a + Q6a + Q7a + Q8a + Q9a)} \end{aligned} \tag{1}$$

The indicator, average HFIAS Score, was calculated using the household scores calculated from Eq. 1.

Average HFIAS Score:

$$\frac{Sum\ (HFIAS\ Scores\ in\ the\ sample)}{\text{Number of HFIAS Scores (i.e. house holds) in the sample}} \tag{2}$$

Thematic content analysis (Creswell & Creswell, 2018) was used to analyze the FGDs and KIIs. The transcripts were manually coded and

Table 2 Categories of food insecurity

Question	*Frequency*		
	Rarely	*Sometimes*	*Often*
1. In the past four weeks, did you worry that your household would not have enough food?	Food secure	Mildly food insecure	
2. In the past four weeks, were you or any household member not able to eat the kinds of foods you preferred because of a lack of resources?			
3. In the past four weeks, did you or any household member have to eat a limited variety of foods due to a lack of resources?		Moderately food insecure	
4. In the past four weeks, did you or any household member have to eat some foods that you really did not want to eat because of a lack of resources to obtain other types of food?			
5. In the past four weeks, did you or any household member have to eat a smaller meal than you felt you needed because there was not enough food?			
6. In the past four weeks, did you or any household member have to eat fewer meals in a day because there was not enough food?			
7. In the past four weeks, was there ever no food to eat of any kind in your household because of lack of resources to get food?	Severely food insecure		
8. In the past four weeks, did you or any household member go to sleep at night hungry because there was not enough food?			
9. In the past four weeks, did you or any household member go a whole day and night without eating anything because there was not enough food?			

Source Modified from Coates, Swindale, and Bilinsky (2007, p. 19)

indexed to identify recurrent and interconnected themes (Krippendorff, 2018). To emphasize stories narrated by the study participants, narrative analysis in the form of quotations was used to substantiate claims and interconnections between themes (Ahmed & Kuusaana, 2021; Bryman, 2015).

5 Results and Discussions

5.1 *Manifestation of Climate Variability and Extremes in the Talensi District*

The findings revealed that the most impactful and prevalent manifestations of climate variability and extremes in the Talensi district are drought (32.5%), high temperatures (30.3%), rainfall variability (13.4), floods (12.5%), and land degradation (11.3%). This finding is in line with Overland (2021) findings that Africa merits a central place in global climate research due to the scale and severity of observed climate impacts, mainly drought and flood. The analysis showed that communities located on high grounds experience intense drought than low-lying communities that mostly experience flooding. A key informant indicated:

> Every year, the same communities get flooded. We live along the White Volta and anytime there is spillage of the Bagre dam which is on the White Volta in Burkina Faso, our community is inundated with severe destruction to our houses and farmlands. Also, anytime heavy rainfalls follow severe drought, flooding is constant here. We are constantly at the mercy of the weather. (key informant, Pwalugu)

The focus group discussions revealed that drought and high temperatures negatively affect agricultural activities. A farmer stated:

> Right after the rainy season, our water bodies including our dams quickly dry up. This increases soil aridity and our lands become dry and difficult to cultivate. We put in much effort and get fewer results. It is frustrating, to say the least. (farmer, Santeng)

Reduced crop yield was experienced as an impact of climate variability. The focus group discussants mentioned frequent and intense droughts, and rising temperatures coupled with rainfall variability. A discussant indicated:

> Growing up in the late 1970s, we used to start planting latest in April. However, now the rains come late and cease early. In some planting seasons, we can wait for as long as June/July to start planting. This 'unfamiliar and unknown' rainfall pattern keeps worsening each season and leading to prolonged dry seasons with its negative impacts on our agriculture activities. (focus group discussant, Winkogo)

Asare-Nuamah (2021) found that in the Southern part of Ghana, climate variability affects agriculture activities through floods, droughts, soil erosion, soil infertility, and soil aridity. Others included changes in farming season, pest invasion, growth of weeds, water availability, loss of livestock, reduction in crop yields, increase in agriculture expenditure, and reduction in quality of meat. These increasingly impact food security primarily through its impact on agriculture which is sensitive to climate and its primary role in determining food supply and providing livelihoods for the rural poor (Holleman et al., 2020). Similar findings emerged in this study conducted in the northern part of Ghana. This points to the pervasive and devastating impacts of climate variability and extremes on food security throughout Ghana.

In terms of land degradation, it was observed through community transact walks in the form of observations and key informant interviews that in addition to floods from excessive rainfall and its associated havoc, high-intensity rainfall has increased erosion and run-off in the communities and thereby affecting agriculture activities. Here, topsoils are mostly washed off and making soils infertile. In addition to land degradation, it was also found that access to land (see WFP et al., 2012; Murken & Gornott, 2022) is important for the households due to their crop-livestock production systems. However, widespread scarcity of agricultural land mainly from land tenure arrangement, deforestation, land conflict, indiscriminate bush burning, and population growth constrained agricultural production and incomes to access food. Also, the changing climate was found to have affected livestock production. They are affected both by drought and excessive rainfall. This is in line with Asare-Nuamah (2021) findings that pests and disease infestations in livestock are rampant in rural Ghana. Also, during drought, water becomes a big problem for livestock as there is less availability of water for them. Moreover, since there is scarcity of pasture during drought, livestock are mostly stunted and die. A key informant commented:

> Due to the ever-changing nature of our climate, during the dry season, our livestock do not have sufficient pasture and water to feed on and most of them do not grow well and die. We move down south with our cattle in search of pasture and water leaving our goats and sheep to be taken care of by our wives and children. Most of them, especially the young ones die. When there are heavy rains too, pests and disease, and drowning are rampant among our livestock. This is a big problem for us. (key informant, Yameriga)

These findings corroborate that of Masekoameng and Maliwichi (2014) that recurring extreme weather events undermine production. Rainfall extremes are detrimental to crop production under rain-fed conditions. Also, altered weather patterns increase crop vulnerability to pest infestation. Droughts followed by intense rains reduce the capacity of water infiltration and lead to flooding. Rainfall extremes have long-term negative impacts on household livelihood. This leads to chronic and acute food insecurity. Also, households' vulnerability to food insecurity rises during protracted drought through the progressive depletion of capital assets and food stocks (Adane et al., 2015) particularly, where there is limited capacity to adapt as is the case in the Talensi district.

5.2 *Measuring Household Food Security*

Climate variability and extremes threaten the production and distribution of food, endanger food accessibility by destabilizing food prices and harm people's health by affecting their nutrition and diets and the quality and quantity of food produced (FAO, 2013). While Table 3 shows the occurrence and frequency-of-occurrence of the HFIAS indicators, Table 4 details the average HFIAS Scores of the food (in)security status of the crop-livestock farmer households (Table 3).

As shown in Table 3, the first indicator of HFIAS reflects anxiety and uncertainty, indicators 2–4 reflect insufficient quality of food, and indicators 5–9 reflect inadequate food intake and its physical consequences

Table 3 HFIAS distribution

Question	*Occurrence*		*Frequency-of-Occurrence*		
	Yes (%)	*No (%)*	*Rarely (%)*	*Sometimes (%)*	*Often (%)*
1	**87.3**	**12.7**	**10.2**	**36.5**	**53.3**
2	**76.3**	**23.7**	**32.3**	**17.7**	**50**
3	**83.2**	**16.8**	**28.1**	**23.2**	**48.7**
4	**78**	**22**	**5.8**	**15.1**	**79.1**
5	**80.1**	**19.9**	**11.9**	**5.2**	**82.9**
6	**72.8**	**27.2**	**12.6**	**5.8**	**81.6**
7	**39.1**	**60.9**	**12.4**	**7.2**	**80.4**
8	**37**	**63**	**14.7**	**6.3**	**79**
9	**19**	**81**	**12.5**	**7.3**	**80.2**

Table 4 Household food in(security) status

Food In(security) Status	*Average HFIAS Scores (%)*
Severely food insecure (1)	42.2
Moderately food insecure (2)	25.2
Mildly food insecure (3)	32
Food Secure (4)	10.2
Food Insecure (1 + 2)	67.4

(Coates et al., 2007). Interestingly, indicators 4–9 were used as food shortage coping strategies by the households. The engagements point to the fact that households either eat less, eat whatever food is available, or do not eat at all in times of food scarcity. The study found that household heads mostly sacrifice available food so their children and other younger family members can eat. Increased reliance on coping strategies is associated with lower food availability (Icheria, 2015). From Table 3, as high as 87.3% of households reported anxiety of food insecurity. Food anxiety is typical and prevalent in food insecure communities (Coates et al., 2007). In as much as these findings confirm that of the WFP, GSS, and MoFA (2012) that the Talensi district has the highest proportion of food insecure households in the Upper East region, relatively fewer households reported severe forms of food insecurity. Here, 39.1% did not eat for a whole day, 37% of went to sleep at night hungry, and only 19% did not eat whole day and night. However, the frequency of their occurrence similar to the other indicators indicates food insecurity among the households.

The above is evident in Table 4 where only 10.2% of the households are food secure with 32% mildly food insecure. Following WFP et al. (2020), food insecurity was calculated by summing up moderately food insecure and severely food insecure households. Thus, in terms of food insecure, as high as 67.4% of the crop-livestock farmer households are food insecure; 25.2% moderately food insecure and 42.2% severely food insecure. Generally, the food insecurity situation in the district was found to be caused by limited household resources, particularly incomes to improve production and access food (food purchases), seasonality of food availability from seasonal changes, and its associated high prices and poor agricultural performance as the crop-livestock production systems in the district continue to remain highly rain-fed, traditional, informal and subsistence-based, and reliant on family labor with limited use of high-yielding seed and improved technology. According to the Environmental Protection Agency (EPA) (2020) and the World Bank (2021), these factors make the agriculture sector highly vulnerable to climate variability and change and characterized by low productivity levels.

Further analysis revealed that 72.3% of the households experience food shortages in the next three months immediately after the harvesting period. This confirms Ahn et al. (2022) conclusion that households that experience food shortages for three months or more with scarce availability and access are classified as severely food insecure. Households that are moderately food insecure experience food shortages for three months with difficulty procuring quality nutritious food. Conversely, mildly food insecure households have food shortages of two months or fewer. Krishnamurthy, Lewis, and Choularton (2012) asserted that in food insecure regions, many smallholder farmers consume their produce and sell it for income in local markets. When production decreases as a result of climate variations, these smallholders (farmers) have less produce to sell for food purchases and also less of their own food production for consumption.

Livestock serve as a buffer during hardship and farmers disinvest them to buy food (Regassa & Stoecker, 2012). In the study communities, it was found that during difficult times, households mostly reduce the size of their livestock by selling them alongside other productive assets for food purchases. This confirms the findings that the Talensi district has much higher levels of poverty (based on wealth index scores) along with very high reliance on cash purchases for food (GSS, 2018; MoFA, 2019; WFP et al., 2012). While this helps the households in the short term, it has long-term implications, if not managed well. It is cautioned that the continuous selling of productive assets such as livestock, poultry, and land dwindles farmers' tangible assets and can expose households to chronic livelihood and food insecurity (Aniah et al., 2019; Antwi-Agyei & Nyantakyi-Frimpong, 2021). A farmer narrated in a focus group discussion:

> During the long dry season, our food stock depletes. It becomes very difficult to buy food because we do not have money. We sell our livestock and other assets mostly at cheaper prices to meet our food needs. There are instances where some households sell all their livestock especially when the rains delay. This is not good, but we have no options since we cannot postpone our hunger. (focus group discussant, Shiega)

The study findings corroborate other studies that climate variability and extremes reduce crop yields and marine foods, increase the spread of pests and diseases, heat stress, result in water and forage challenges in livestock, reduce growth and reproduction and loss of livestock, and reduce milk and wool production (Dasgupta et al., 2017; Thornton & Herrero, 2015). Consequently, these affect production and increase food prices and affect the purchasing power and the ability of households to access sufficient, safe, nutritious, and preferred food (Asare-Nuamah, 2021; Hasegawa et al., 2015; Wossen et al., 2018). Therefore, for crop-livestock smallholder farmers in the Talensi district to effectively adapt to climate variability and extremes consistent and achieve food security, there is the need to build their adaptive capacity.

6 Conclusion and Policy Implications

The study findings show evidence of how climate variability and extremes are negatively impacting agri-food systems and acting as key drivers of food insecurity among crop-livestock smallholder farmers' households in the Talensi district. It was found that only 10.2% of the crop-livestock farmer households are food secure. Thirty-two percent (32%) are mildly food insecure and 67.4% food insecure (25.2% moderately food insecure and 42.2% severely food insecure). Due to the highly exposed nature and sensitivity of crop-livestock smallholder farmer households' production systems to climate variability and extremes, the distribution of the HFIAS occurrence and frequency-of-occurrence as detailed in Tables 3 and 4 showed food insecurity among the households. The study findings lend credence to the fact that it is not only extreme climate events including droughts and floods that negatively affect agricultural systems and result in negative impacts on livelihoods, food prices, income, and coping capacities but also rainfall variability in the form of changing seasonal patterns; shorter rainy seasons and poorly distributed rainfall within seasons (Holleman et al., 2020), land degradation, and pest and diseases infestation. The farmer households adopt indicators 4 to 9 (see Table 2) as their food shortage coping strategies. While indicator 4 reflects insufficient quality of food, indicators 5 to 9 reflect inadequate food intake and its physical consequences. This falls in line with the conclusion that the strongest impacts of increasing climate variability and extremes are felt on food availability with the overall fallout far more complex, and significantly undermining food access (Holleman et al., 2020) as revealed by the HFIAS. The onus of responsibility, therefore, lies on all levels of government, policymakers, and implementers to prioritize food security interventions among crop-livestock smallholder farmer households.

A number of policy interventions to address climate variability and household food security in the Talensi district are, therefore, being implemented. Two such major interventions are the Landscape Restoration and Small-Scale Mining and the Creating Lands of Opportunity: Transforming Livelihoods through Landscape Restoration in the Sahel Project. The Ghana Landscape Restoration and Small-Scale Mining Project seek to strengthen integrated natural resource management and increase benefits to its beneficiary communities through institutional and regulatory strengthening for participatory landscape management and sustainable crop and forest landscape management and formalize and ensure accountability of artisanal and small-scale mining (Ministry of Land and Natural Resources [MLNR], 2021). It was found that the project is supporting farmers restore degraded lands for agricultural productivity, strengthening sustainable management of forest landscapes to conserve biodiversity and ecosystem services, and also land use planning for integrated landscape management to optimize land use to land characteristics. Also, the district is benefiting from the Creating Lands of Opportunity: Transforming Livelihoods through Landscape Restoration in the Sahel Project. Here, the crop-livestock farmers were found to benefit from interventions to improve their production and economic opportunities for improved livelihood. Under the project, farmers' capacities are being built on improved and environmentally friendly farming practices.

To improve and build on the gains from these interventions, the chapter calls for diversification of livelihood and income sources that are less sensitive to climate variability and extremes to increase farmers' adaptive capacity and improve their purchasing power. According to Wossen and Berger (2015) and Ayinu et al. (2022), policy interventions of this nature should be implemented along with measures to reduce reliance on rain-fed agriculture through the transformation of production systems. Thus, a fundamental transformation is required and not just simply 'tweaking' existing agricultural systems. A regenerative agricultural system consisting of a mosaic of sustainable production methods that marry local knowledge with modern techniques is critical (United Nations Conference on Trade and Development [UNCTAD], 2010; Abegunde et al., 2022). In line with Amoah and Simatele (2021), respondents' prior knowledge and awareness of climate variability and extremes should be leveraged to build their adaptive capacities.

Due to the environmentally fragile dryland (non-irrigated) area of the Talensi district, declining rainfall, disappearing forest cover and associated biodiversity, and loss of productivity from declining soil fertility (Kandel et al., 2021; WVA, 2018), this study calls for major public and private sector investments in irrigation, efficient water management, and water resources development such as water harvesting and saving technologies, wastewater management, and new water sources to increase all year food production. Also, because of erratic rainfall, prolonged droughts and floods that are already affecting productivity and disrupting households' livelihoods, climate-smart agriculture (CSA) strategies and conservation agriculture practices are fundamental to transforming agricultural systems and providing resilience pathways (Diko et al., 2021). These include integrated and improved crop-livestock management; diet improvement for livestock, planting drought, and heat tolerant crops; use and efficient organic manure management, crop diversification, and minimum tillage among others. These emerging CSA technologies and practices (see Zougmoré et al., 2018; Abegunde et al., 2022; Hashmiu et al., 2022) provide ways to mobilize political support, promote sustainable agriculture production, improve farm productivity, increase farmers' income, improve rural livelihoods, and build farmers adaptive and resilience capacities and production systems (Diko et al., 2021; Zougmoré et al., 2018). Again, investments in transportation networks and conditions, market access and agricultural research, extension, and other methods to reduce food production losses (Cudjoe et al., 2021) are critical. To make agriculture attractive, especially at the smallholder level, this paper agrees with Abegunde et al. (2022) to use extension services for information dissemination in more attractive and productive ways to rebrand agriculture as a lucrative venture and not a 'lender of last resort' for those without options.

In response to restoring degraded lands, the study calls for sustainable natural regeneration practices that can enhance livelihood and food security. This is critical as natural regeneration is relatively cheap and simple, and has higher tree survival rates as the species that regenerate naturally are locally adapted and significantly reduce the cost of land restoration (FAO, 2019; Lohbeck et al., 2020). One of such practices is Farmer Managed Natural Regeneration (FMNR). FMNR is a full rural development model that combines natural regeneration with tree-dependent enterprises such as apiculture, livestock production, and local value chains for wood, timber, and other non-timber forest products (Nakyeyune et al., 2018). This integrated approach contributes to increased crop yields, increased incomes, improved livestock production, diversified food options, and improved access to food and nutrition (Rinaudo et al., 2019; WVA, 2018). The study agrees with Adiku et al. (2017) that with the increasing climate variability and extremes, agricultural insurance is an important tool to manage climate-related shocks. However, while major steps to promote agricultural insurance are evolving in Africa, the study calls for the implementation of the Savings for Transformation (S4T) model. The model brings individuals together in cycles of saving and borrowing credit for their business and farm initiatives and/or family needs. It is inexpensive, community-led and creates access to financial credit with lower interest rates. It builds a culture of savings and creates access to microcredit for business start-ups and investments in agriculture (Weston & Hong, 2013) to improve livelihoods and food security.

References

Abegunde, V. O., Sibanda, M., & Obi, A. (2022). Effect of climate-smart agriculture on household food security in small-scale production systems: A micro-level analysis from South Africa, *Cogent Social Sciences*, *8*(1). https://doi.org/10.1080/23311886.2022.2086343

Adane, D. M., Atnafe, A. D., & Ahmed, H. M. (2015). The status of food availability in the face of climate change and variability in Choke Mountain Watersheds, Central Ethiopia. *Journal of Development and Agricultural Economics, 7*(10), 358–372. https://doi.org/10.5897/JDAE2015.0672

Adiku, S. G. et al. (2017). *Weather-index based crop insurance as a social adaptation to climate change and variability in the Upper West Region of Ghana: Developing a participatory approach.* CGIAR Research Program on Climate Change, Agriculture and Food Security (CCAFS).

Ahmed, A., & Kuusaana, E. D. (2021). Cattle ranching and farmer-herder conflicts in sub-Saharan Africa: Exploring the conditions for successes and failures in Northern Ghana. *African Security, 14*(2), 132–155. https://doi.org/10.1080/19392206.2021.1955496

Ahn, J. et al. (2022). Food security and agricultural challenges in West-African rural communities: A machine learning analysis. *International Journal of Food Properties, 25*(1), 827–844.

Alemu, T. and Mengistu, A. (2019). Impacts of climate change on food security in Ethiopia: Adaptation and mitigation options: A review. In Castro, P.,

Azul, A., Filho, W. L. and U. Azeiteiro (Eds.), *Climate change-resilient agriculture and agroforestry*. Springer, Cham. https://doi.org/10.1007/978-3-319-75004-0_23.

Ali, S. et al. (2017). Climate change and Its impact on the yield of major food crops: Evidence from Pakistan, *Foods, 6*(6). https://doi.org/10.3390/foods6060039.

Amoah, L. N. and Simatele, M. D. (2021). Food security and coping strategies of rural household livelihoods to climate change in the Eastern Cape of South Africa, *Frontiers in Sustainable Food Systems, 5*(692185). https://doi.org/10.3389/fsufs.2021.692185.

Aniah, P., Kaunza-Nu-Dem, M. K., & Ayembilla, J. A. (2019). Smallholder farmers livelihood adaptation to climate variability and ecological changes in the savanna agro ecological zone of Ghana. *Heliyon, 5*, e01492. https://doi.org/10.1016/j.heliyon.2019.e01492

Antwi-Agyei, P., & Nyantakyi-Frimpong, H. (2021). Evidence of climate change coping and adaptation practices by smallholder farmers in Northern Ghana. *Sustainability, 13*(3), 1308. https://doi.org/10.3390/su13031308

Antwi-Agyei, P., Stringer, L. C., & Dougill, A. J. (2014). Livelihood adaptations to climate variability: Insights from farming households in Ghana. *Regional Environmental Change, 14*, 1615–1626. https://doi.org/10.1007/s10113-014-0597-9

Arribas, A. et al. (2011). The gloSea4 ensemble prediction system for seasonal forecasting. *Monthly Weather Review, 139*(6), 1891–1910. https://doi.org/10.1175/2010MWR3615.1

Asare-Nuamah, P. (2021). Climate variability, subsistence agriculture and household food security in rural Ghana. *Heliyon, 7*(4), e06928. https://doi.org/10.1016/j.heliyon.2021.e06928

Awazi, N. P. et al. (2022). Climate-Smart and agro-ecological farming systems of smallholder farmers' in Galanakis, C. M. (ed.), *Environment and climate-smart food production*. Springer, Cham. https://doi.org/10.1007/978-3-030-71571-7_2.

Ayinu Y. T. et al. (2022). Impact of climate variability on household food security in Godere District, Gambella Region, Ethiopia. *Climate Services, 27*(100307). https://doi.org/10.1016/j.cliser.2022.100307.

Baldos, U. L., & Hertel, T. W. (2014). Global food security in 2050: The role of agricultural productivity and climate change. *The Australian Journal of Agricultural and Resource Economics, 58*(4), 554–570. https://doi.org/10.1111/1467-8489.12048

Belloumi, M. (2014) *Investigating the linkage between climate variables and food security in ESA countries*. International Food Policy Research Institute.

Bryman, A. (2015) *Social research methods*. 5th edn. Oxford University Press.

Callo-Concha, D. (2018). Farmer perceptions and climate change adaptation in the West Africa Sudan Savannah: Reality check in Dassari, Benin, and Dano, Burkina Faso. *Climate, 6*(2), 44. https://doi.org/10.3390/cli6020044

Coates, J., Swindale, A. & Bilinsky, P. (2007). *Household food insecurity access scale (HFIAS) for measurement of food access: Indicator guide (Version 3).* Food and Nutrition Technical Assistance III Project (FANTA).

Creswell, J. W. (2014) *Research design: qualitative, quantitative, and mixed methods approaches*. 4th edn. Sage Publications.

Creswell, J. W. and Creswell, J. D. (2018). *Research design: Qualitative, quantitative, and mixed methods approaches.* 5th edn. Sage Publishing.

Cudjoe, G. P., Antwi-Agyei, P. & Gyampoh, B. A. (2021). The effect of climate variability on maize production in the Ejura-Sekyedumase Municipality, Ghana, *Climate*, *9*(145). https://doi.org/10.3390/cli9100145

Dasgupta, S. et al. (2017). The impact of aquatic salinization on fish habitats and poor communities in a changing climate: Evidence from southwest coastal Bangladesh. *Ecological Economics, 139*, 128–139. https://doi.org/10.1016/j.ecolecon.2017.04.009

Diko, S. K. et al. (2021). Are local development plans mainstreaming climate-smart agriculture? A mixed-content analysis of medium-term development plans in semi-arid Ghana. *Socio-Ecological Practice Research, 3*, 185–206. https://doi.org/10.1007/s42532-021-00079-2

EPA. (2020). *Ghana's fourth national communication to the United Nations framework convention on climate change*. EPA.

FAO. et al. (2017). *The state of food security and nutrition in the World 2017. Building resilience for peace and food security.* FAO.

FAO. et al. (2018). *The state of food security and nutrition in the World 2018. Building climate resilience for food security and nutrition.* FAO.

FAO. (2021). The state of food security and nutrition in the world, et al. (2021). Transforming food systems for food security, improved nutrition and affordable healthy diets for all. *FAO.* https://doi.org/10.4060/cb4474en

FAO, IFAD, WFP. (2014). *The state of food insecurity in the World 2014. Strengthening the enabling environment for food security and nutrition.* FAO.

FAO. (2013). *Food security and agricultural mitigation in developing countries: Options for capturing synergies.* FAO.

FAO. (2015). *Climate change and food systems: Global assessments and implications for food security and trade.* FAO.

FAO. (2019). *Restoring forest landscapes through assisted natural regeneration (ANR)—A practical manual.* FAO.

Firdaus, R. B. et al. (2019). Does climate change only affect food availability? What else matters? *Cogent Food & Agriculture, 5*(1). https://doi.org/10.1080/23311932.2019.1707607.

GSS. (2014). *2010 population and housing census: Talensi District. District analytical report*. GSS.

GSS. (2018). *Ghana Multiple Indicator Cluster Survey (MICS 6), 2017/18*. GSS.

Hasegawa, T. et al. (2015). Consequence of climate mitigation on risk of hunger. *Environmental Science & Technology, 49*(12), 7245–7253. https://doi.org/10.1021/es5051748

Hashmiu, I., Agbenyega, O. and Dawoe, E. (2022). Cash crops and food security: Evidence from smallholder cocoa and cashew farmers in Ghana, *Agriculture & Food Security 11*(12). https://doi.org/10.1186/s40066-022-00355-8

Holleman, C. et al. (2020). *The impact of climate variability and extremes on agriculture and food security—An analysis of the evidence and case studies*. FAO. https://doi.org/10.4060/cb2415en

Icheria, B. K. (2015). Household food insecurity and coping strategies among small scale farmers in Tharaka Central Division, Kenya. *International Journal of Humanities Social Sciences and Education (IJHSSE), 2*(2), 63–76.

INDDEX Project. (2018). *Data4Diets: Building blocks for diet-related food security analysis*. Tufts University.

IPCC. (2012). *Managing the risks of extreme events and disasters to advance climate*. Cambridge University Press.

IPCC. (2014). *Climate change 2014: Impacts, adaptation and vulnerability*. Cambridge University Press.

IPCC. (2018). Summary for Policymakers in *Global Warming of 1.5°C*. IPCC.

IPCC. (2019). *Climate change and land: an IPCC special report on climate change, desertification, land degradation, sustainable land management, food security, and greenhouse gas fluxes in terrestrial ecosystems*. IPCC

IPCC. (2022). *Summary for policymakers: Climate change 2022, impacts, adaptation and vulnberbility*. IPCC.

Kandel, M. et al. (2021). Assessing social equity in farmer-managed natural regeneration (FMNR) interventions. *Ecological Restoration, 29*(1–2), 64–76. https://doi.org/10.3368/er.39.1-2.64

Krippendorff, K. (2018). *Content analysis: An introduction to its methodology*. 4th edn. Sage Publishing.

Krishnamurthy, P. K., Lewis, K., & Choularton, R. J. (2012). *Climate impacts on food security and nutrition. A review of existing knowledge*. WFP.

Kumasi, T. C., Antwi-Agyei, P., & Obiri-Danso, K. (2019). Small-holder farmers' climate change adaptation practices in the Upper East Region of Ghana. *Environment, Development and Sustainability, 21*, 745–762. https://doi.org/10.1007/s10668-017-0062-2

Lohbeck, M. et al. (2020). Drivers of farmer-managed natural regeneration in the Sahel. Lessons for restoration, *Scientific Reports*, 10(15038). https://doi.org/10.1038/s41598-020-70746-z.

Masekoameng, M., & Maliwichi, L. L. (2014). Determinants of food accessibility of the rural households in Sekhukhune District Limpopo Province, South Africa. *Journal of Human Ecology, 47*(3), 275–283. https://doi.org/10.1080/09709274.2014.11906762

MLNR. (2021). Ghana landscape restoration and small-scale mining project: Environmental and Social Management Framework (ESMF). MLNR.

MoFA. (2019). *Agriculture in Ghana. facts and figures.* MoFA.

Murken, L. and Gornott, C. (2022). The importance of different land tenure systems for farmers' response to climate change: A systematic review *Climate Risk Management, 35*(100419). https://doi.org/10.1016/j.crm.2022.100419.

Nakyeyune, C., Okia, C. A., & Kiyingi, G. (2018). *Farmer managed natural regeneration: A user's guide for practitioners in Uganda*. World Agroforestry Centre (ICRAF).

Overland, I. et al. (2021). Funding flows for climate change research on Africa: Where do they come from and where do they go? *Climate and Development*. https://doi.org/10.1080/17565529.2021.1976609

Phiiri, G. K., Egeru, A., & Ekwamu, A. (2016). Climate change and agriculture nexus in sub-saharan Africa: The agonizing reality for smallholder farmers. *International Journal of Current Research and Review, 8*(2), 57–64.

Poppy, G. et al. (2014). Achieving food and environmental security: New approaches to close the gap, *Philosophical Transactions of the Royal Society of London B: Biological Sciences, 369*(1639). https://doi.org/10.1098/rstb.2012.0272.

Regassa, N., & Stoecker, B. J. (2012). Household food insecurity and hunger among households in Sidama district, southern Ethiopia. *Public Health Nutrition, 15*(7), 1276–1283. https://doi.org/10.1017/S1368980011003119

Rinaudo, T., Muller, A., and Morris, M. (2019). *Farmer managed natural regeneration (FMNR) manual.* WVA.

Slovin, E. (1960). Slovin's formula for sampling Technique. https://prudencexd.weebly.com/

Stephanie, E. (2003). Slovin's formula sampling techniques. Houghton-Mifflin.

Tangonyire, D. F. and Akuriba, G. A. (2021). Socioeconomic factors influencing farmers' specific adaptive strategies to climate change in Talensi district of the Upper East Region of Ghana, *Ecofeminism and Climate Change, 2*(2), 50–68.https://doi.org/10.1108/EFCC-04-2020-0009.

Thornton, P. K., & Herrero, M. (2015). Adapting to climate change in the mixed crop and livestock farming systems in sub-saharan Africa. *Nature Climate Change, 5*, 830–836. https://doi.org/10.1038/nclimate2754

UNCTAD. (2010). *Creative economy report 2010.* UNCTAD.

UNDP. (2018). *Climate change adaptation in Africa: UNDP synthesis of experiences and recommendations.* UNDP.

UNEP. (2021). *Adaptation gap report 2020*. UNEP.

UNFCCC. (2021). *Nationally determined contributions under the Paris agreement: Synthesis report*. UNFCC.

USAID. (2013). *Food security in Kenya: Situation report*. USAID.

Weldearegay, S. K. and Tedla, D. G. (2018). Impact of climate variability on household food availability in Tigray, Ethiopia, *Agriculture & Food Security, 7*(6). https://doi.org/10.1186/s40066-017-0154-0.

Weston, P. and Hong, R. (2013). *Social return on investment report*. WVA.

WFP, GSS and MoFA. (2012). *Comprehensive food security & vulnerability analysis. Ghana 2012*. GoG.

WFP. et al. (2020). *Comprehensive food security & vulnerability analysis. Ghana 2020*. GoG.

World Bank. (2021). *Climate risk profile: Ghana*. World Bank.

Wossen, T., & Berger, T. (2015). Climate variability, food security and poverty: Agent-based assessment of policy options for farm households in Northern Ghana. *Environmental Science & Policy, 47*, 95–107. https://doi.org/10.1016/j.envsci.2014.11.009

Wossen, T. et al. (2018). Impacts of climate variability and food price volatility on household income and food security of farm households in East and West Africa. *Agricultural Systems, 163*, 7–15. https://doi.org/10.1016/j.agsy.2017.02.006

WVA. (2018). *Farmer managed natural regeneration: Contributing to reduced household hunger in the Talensi District, Ghana*. WVA.

Zougmoré, R. B. et al. (2018). Facing climate variability in sub-saharan Africa: Analysis of climate-smart agriculture opportunities to manage climate-related risks, *Cahiers Agricultures, 27*(3). https://doi.org/10.1051/cagri/2018019.

CHAPTER 11

Putting People First in Climate Governance: The Role of South Africa's Indigenous Values in Securing Ecologically Sustainable Development

Felix Dube and Aubrey Manthwa

1 Introduction

Indigenous peoples have an immense contribution to make in addressing climate change risks and hazards through indigenous sustainability, which entails the use of indigenous knowledge systems to carefully balance the needs of humanity and ecological preservation for the benefit of the present generation and posterity (Tsosie, 2019). Hence,

F. Dube (✉) · A. Manthwa
Department of Public, Constitutional and International Law, College of Law, University of South Africa, Pretoria, South Africa
e-mail: felix@fdube.co.za

A. Manthwa
e-mail: manthat@unisa.ac.za

M. Addaney et al. (eds.), *Climate Change in Africa*, https://doi.org/10.1007/978-3-031-30050-9_11

there is a growing acceptance among scholars that indigenous knowledge systems and values have a crucial role to play in maximising the socio-economic well-being of indigenous communities through ecologically sustainable development. This is also because of the vulnerability of indigenous communities to climate change, as seen in increasing instances of droughts, flooding and drastic temperature changes (Jerez, 2021). Given the connection between climate change and sustainable development (Addaney & Cobbinah, 2019), this chapter analyses how indigenous values in South Africa could influence the understanding, formulation and implementation of policies and legislation for securing ecologically sustainable development. In this context, this chapter argues that it is necessary to consider relevant indigenous values and principles that command a strong presence in South Africa's environmental policy and regulatory approaches. Such values and principles may be the ideal premises from which to mitigate climate change by, *inter alia*, securing ecologically sustainable development, as required by Section 24 of the Constitution of the Republic of South Africa, 1996 (the Constitution).

We identify *batho pele* the indigenous value of putting people first in decisions that affect them, including decisions on the exploration of the main drivers of climate change—oil and gas—as one of the ideal premises from which to mitigate and adapt to climate change through ecologically sustainable development. We submit that when properly implemented in constitutional environmental law *batho pele* could become an essential component of balancing the competing interests of protecting the climate and economic and social development. Our approach in this regard is informed by the community-centred aspects of *batho pele* and its recognition as a policy imperative in South Africa's governance and service delivery framework, including climate change mitigation and adaptation. We begin the discussion with an exposition of ecologically sustainable development within the broad scope of international environmental law discourse to contextualise its inclusion and framing in the South African Constitution. We then examine the potential of indigenous values as instruments for mitigating climate change through securing ecologically sustainable development. This lays the foundation for our analysis of *batho pele* as an indigenous value that should enjoy more policy and legislative recognition in South Africa's climate and environmental governance framework.

2 Ecologically Sustainable Development and the South African Constitution

Ecologically sustainable development is one of many concepts in the discourse on mitigating and adapting to climate change. In the *BP* case, the court agreed that, in its essence, ecologically sustainable development is about the following:

a. preservation of natural systems so that future generations can benefit from them;
b. sustainable use of natural resources;
c. equitable use of natural resources; and
d. integration of environmental consideration into economic, financial and social development.

When put into practice, these definitional elements of ecologically sustainable development secure clean air, protect vital plant nutrients and ecosystem services and ultimately mitigate climate change. It is in this context that Bosselmann & Kim (2013, p. 307) define ecologically sustainable development as 'development that meets the needs of the present while safeguarding Earth's life-support system, on which the welfare of current and future generations depends.' To secure ecologically sustainable development, states need to develop institutions, processes and laws to safeguard the environment from damage and overexploitation (Marten, 2001, p. 157). In line with this requirement, South African law constitutionalises the obligation of the state to secure ecologically sustainable development in Section 24(b)(iii) of the Constitution. This section bestows every person who lives in South Africa a right 'to have the environment protected, for the benefit of present and future generations, through reasonable legislative and other measures that ***secure ecologically sustainable development*** and use of natural resources while promoting justifiable economic and social development' (our emphasis).

The constitutional duty in Section 24(b) of the South African Constitution falls within the broad scope of what the High Court has considered part of the corresponding right to sustainable development in the *Baleni* case. This is because while the South African Constitution does not bestow an express right to sustainable development, Section 24 expressly mentions the tripartite definitional elements of sustainable development—environmental protection, economic development and social protection, leading the court to conclude that the section confers a right to sustainable development. However, South African courts have not engaged in depth with the duty to secure ecologically sustainable development in depth. A search of the case law reveals numerous judicial references to ecologically sustainable development but only in passing and often as part of quotations from Section 24 of the Constitution. Since the judiciary plays a pivotal role in the protection of the environment (Mwanza, 2019, 2022), its jurisprudence on ecologically sustainable development would have been invaluable for this chapter. The difficulty is compounded by scholarly approaches that do not view securing ecologically sustainable development as a right but as a qualifier and constitutional value that informs the interpretation of the right to have the environment protected (Chikuruwo, 2022).

Like all constitutional duties, securing ecologically sustainable development is justiciable, giving the courts powers to test the rationality, proportionality and reasonableness of legislation and executive decisions that impinge on it. This is a constitutionality test in that in the *Fuel Retailers* case, it was emphasised that when courts review decisions about ecologically sustainable development, the issue is not whether development is financially and economically justified but whether it is balanced with its negative impacts on the environment. This also considers intergenerational equity. In this regard, the courts have noted that when the constitutional drafters elevated the protection of the environment into a justiciable right, they set South Africa on a path to attain environmental protection in a manner that seriously considers socio-economic issues that arise from actions dealing with the environment. The *BP* case illustrates this judicial reasoning.

The constitutional duty to secure ecologically sustainable development should also be understood in the context that the Constitution 'is founded on a set of aspirational but foundational values that should permeate all society' (Kotzé, 2016, p. 158). This means that securing ecologically sustainable development, among other constitutional duties, could be considered broadly with reference to constitutional values, some of which derive from indigenous customs, and which are relevant to the present discussion as they are intended to achieve human dignity, equality, and to advance human rights and freedoms—as stipulated in the founding provisions in Chapter 1 of the South African Constitution. They are also relevant in that they form part of the tenets of South Africa's environmental constitutionalism and other attributes of the legal system that protect ecological integrity.

3 Legislative Approaches to Securing Ecologically Sustainable Development

In line with its constitutional obligations to enact legislation and use other measures to secure ecologically sustainable development, South Africa enacted the *National Environmental Management Act* (commonly known as NEMA). The courts have confirmed in the *HTF Developers* case that this statute is the legislation which the Constitution envisages in Section 24(b). Section 2(4) of NEMA requires the consideration of several factors in the context of sustainable development as envisaged in Section 24(b) of the Constitution. Among these, Section 2(4)(a) of NEMA provides "that negative impacts on the environment and on people's environmental rights be anticipated and prevented, and where they cannot be altogether prevented, are minimised and remedied." There can be no clearer expression of the need for the mitigation of environmental harm than this one.

In light of the requirements of Section 2(4) of NEMA, South African courts emphasise the need to carefully balance decisions about the environment with the rights, interests and values of affected people. A proper balance can be achieved through *batho pele,* i.e., putting people first. We will show that this indigenous value and its principles link to the essentiality of human well-being in economic initiatives which involve the extraction of resources from the natural environment. In the *Fuel Retailers* case, the court recognised that the exploitation of resources for economic development should not affect other constitutional rights and emphasised that:

> Development cannot subsist upon a deteriorating environmental base. Unlimited development is detrimental to the environment and the destruction of the environment is detrimental to development. Promotion of development requires the protection of the environment. Yet the environment cannot be protected if development does not pay attention to the costs of environmental destruction. The environment and development are thus inexorably linked.

Thus, the court showed that the protection of the environment requires sustainable exploitation of resources. It confirmed that South African environmental law should be understood in the context of international environmental law, whose foundational cornerstone is sustainable development (Carruthers, 2001, pp. 93–112). However, present conceptions of sustainable development have their own shortcomings. Despite its prominence in environmental policies, statutes and constitutions across the globe, sustainable development has failed to deliver on its tripartite promises of universal and just economic development, an increase in social well-being and environmental protection (Viñuales, 2013, pp. 3–13). There are many reasons for this failure. One principal consideration is the belief that the idea of sustainable development pursues a skewed type of development, namely a form of neoliberal economic development that masquerades as development that offers economic and social benefits for everyone when in fact, it exploits natural resources as if they are infinite (Springer, 2016). Some critics accordingly consider sustainable development as 'a concept based on false consensus' (Kothari et al., 2014, p. 366) and even a doctrinal palliative (Kotzé, 2016, p. 7).

While we accept that sustainable development is failing to avert environmental challenges facing the modern era, such as climate change, we argue that its inadequacies can be mitigated by roping in indigenous knowledge systems and their values since these do not suffer from the burden of neoliberal alignment and its materialistic underpinnings. In many parts of the world, including South Africa, indigenous values have been proposed as part of the solution to the shortcomings of sustainable development (Van Norren, 2020). In the following section, we lay the theoretical framework for the consideration of indigenous values in the discourse on ecologically sustainable development and how it may contribute to climate change mitigation and adaptation.

4 Indigenous Values as Instruments for Securing Ecologically Sustainable Development: The Case of *Batho Pele*

4.1 Indigenous Values in Context

One of the most remarkable aspects of the modern South African legal system is that it does not only encompass the tenets of constitutionalism that are generally understood and recognised in democracies across the globe but also includes indigenous values. These values are not enshrined in the constitutional text but have become core pillars of constitutional adjudication and policymaking. Constitutionally, they enjoy judicial recognition as unwritten founding constitutional values which serve as interpretative aids and adjudicative tools. An example in this regard is *ubuntu*, which the Constitutional Court used to protect the right to life in the *Makwanyane* case. As we have noted at the beginning, indigenous values have a crucial role to play in securing ecologically sustainable development in South Africa. Their relevance has a historical context in the democratic transition brought into reality that indigenous knowledge systems and their values are relevant *even if* they do not serve the interests of neoliberalism. This contrasts with current conceptions of sustainable development, which Adelman (2013, p. 20) says 'epitomise the epistemological crisis of modernity, in which local, indigenous and traditional knowledge are valid only to the extent that they serve global capitalism.'

It is for this reason that we believe that it may be worthwhile for the South African legal system and its counterparts in the continent to return to their indigenous roots, values and traditions, which could offer better alternatives for the protection of the environment. Indigenous knowledge systems and their values have never been associated with the unsustainable use of natural resources but have, on the contrary, contributed to sustainability through what could be termed as indigenous sustainability (McGregor et al., 2020, pp. 35–40). South Africa has made strides in this regard, starting with the Constitution of 1996, whose adoption made it possible for judges to experiment with indigenous values by incorporating them into South Africa's newly found system of judicial review. In the *Makwanyane* case, the Constitutional Court opened the doors for this experimentation by observing that indigenous value systems are the correct premise from which to build an inclusive society and that they are wholly related to the achievement of freedom and equality.

From a historical perspective, indigenous values are important in this work because colonialism relegated indigenous values to non-entities and gave them no consideration in the enactment of legislation, including legislation on environmental and other issues, as pointed out in the *Merafong* case. As such, indigenous values that have been historically disregarded could be useful for decolonising the law when they are injected into judicial decision-making in the hope of developing inclusive jurisprudence. When it comes to issues of environmental protection, sustainable development and its ecological component, 'the central theme is that in African tradition personal identity cannot be separated from the environment as a whole which encompass the natural, socio-cultural, physical and spiritual environment' (Church, 2012, p. 524). It is for this reason that we propose indigenous knowledge and its values as potential instruments to secure ecologically sustainable development. In the following section, we elaborate on how *batho pele*, an indigenous value, could help South Africa in this regard.

4.2 *Batho Pele: Meaning and Usage in Environmental Discourse*

Batho pele is a Northern Sotho maxim which means 'people first.' In the hope of ensuring that the interests and needs of the people are considered first in government policy and other acts that affect them, South Africa introduced the *White Paper on Transforming Public Service Delivery (Batho Pele)*. In para 1.1.2, the *White Paper* says that it seeks 'to provide a policy framework and a practical implementation strategy for the transformation of public service delivery.' It lists eight principles based on *batho pele*, which must guide the provision of public services and the making of decisions that affect communities: public consultation; setting and maintenance of service standards; enhancing access; courtesy; provision of useful information; increasing openness and transparency in government decisions making; prompt remediation of mistakes and failures; and ensuring value for money. According to the *Van der Merwe v Taylor* case, the essence of *batho pele* is to ensure that public administration 'serve[s] the best interests of the public by enabling the achievement of individual rights encompassed in the provisions of the Constitution.' In the context of the present chapter, this entails the fulfilment of the right to have the environment protected through securing ecologically sustainable development.

Like all organs of State, the Department of Forestry, Fisheries and the Environment, which is responsible for the implementation of environmental legislation and taking administrative decisions such as granting environmental authorisations for oil and gas exploration, embraced *batho pele*. Hence, it began to 'include objectives and activities which relate to the implementation of *batho pele* within strategic and business plans' in the environmental sector. Outlining its vision, the department said that it would improve the monitoring and evaluation of *batho pele* principles in the environmental sector by setting standards for their implementation. This entailed setting environmental outcomes-based targets (which measure the state of the environment) to judge the performance of the sector. This approach was informed by the *White Paper* as well as environmental legislation which requires the state to put the interests of communities first when taking environmental decisions that affect them.

The application of *batho pele* to securing ecologically sustainable development should be understood in the context of Section 2(2) of NEMA, which prescribes that 'environmental management must place people and their needs at the forefront of its concern, and serve their physical, psychological, developmental, cultural and social interests equitably.' Hence, at the heart of this provision is the legislative requirement that when environmental management decisions are made, they must be based on the interests of the people first, as highlighted in the *Fuel Retailers* case:

> One of the key principles of NEMA requires people and their needs to be placed at the forefront of environmental management – *batho pele*. It requires all developments to be socially, economically and environmentally sustainable. Significantly for the present case, it requires that the social, economic and environmental impact of a proposed development be 'considered, assessed and evaluated' and that any decision made 'must be appropriate in the light of such consideration and assessment'. This is underscored by the requirement that decisions must take into account the interests, needs and values of all interested and affected persons.

Thus, NEMA seeks to achieve what the Constitution envisages in Section 24. It requires the government to consider the interests of the people first when making environmental decisions within the context of development. The first consideration that comes into mind when applying *batho pele*, as expressed in the *White Paper* and in Section 2(2) of NEMA, is public consultation. The government is obliged to consult communities when making environmental decisions that could affect their environment, ecologically integrity or any aspect of their lives. This is meant to achieve democratic outcomes in environmental management. It stands in stark contrast to the erstwhile governance system in which poor people, mostly indigenous communities, were not consulted in the making of decisions that adversely affected their environments. Thus, *batho pele* contributes to public participation in the legislative procedure and fits within the broad constitutional vision to ensure a just and inclusive society that promotes social justice, as alluded to in the Constitution's preamble.

In our analysis, the values derived from *batho pele* could be interpreted to advance a right to refuse development that is ecologically unsustainable. This has been demonstrated in a recent case in which the court had to put the people first.

4.3 Batho People (People First) in Climate Change Governance

In addition to constitutional rights and legislation on environmental governance, South Africa has taken several other steps in the direction of mitigating climate change and adapting the country to its negative consequences. These steps include signing the *Paris Agreement* and submitting its National Determined Contributions (South Africa, 2021). In accordance with its obligations under the *Paris Agreement*, South Africa committed to "a fixed target for greenhouse gas emissions levels of 398–510 MtCO2e by 2025, and 350–420 MtCO2e by 2030, compared to 398 and 614 Mt CO2e between 2025 and 2030 as communicated in the first NDC" (United Nations Climate Change, 2021). The Nationally Determined Contributions are informed by the *National Climate Change White Paper*, which stipulates that South Africa's climate change response "will prioritise indigenous knowledge and local adaptative responses." This can be seen as an acknowledgement of the need to put people first in responding to the perils of climate change.

In order to bring its *Climate Change White Paper* into implementation, South Africa adopted a *National Climate Change Adaptation Strategy*, which recognises the importance of indigenous knowledge in climate adaptation and stipulates that such knowledge should be derived from the lived experiences of the indigenous peoples. It also says in implementing it, the government will "Place people at the centre" and that:

> Adaptation action must place people and their needs at the forefront of its concern, and serve their physical, psychological, developmental, cultural and social interests equitably; Special consideration of vulnerable groups, such as women, the elderly, and children shall receive specific attention in adaptation planning and implementation.

At the time of writing, South is in the process of enacting a *Climate Change Bill*. This Bill seeks, *inter alia*, to secure ecologically sustainable development by protecting and preserving "the planet for the benefit of present and future generations of humankind." In clause 3(f), it recognises the specific vulnerabilities of indigenous communities and other exposed groups by acknowledging:

> the need for decision-making to consider the special needs and circumstance of localities and people that are particularly vulnerable to the adverse effects of climate change, including vulnerable workers and groups such as women, especially poor and rural women, children, especially infants and child-headed families, the aged, the poor, the sick and the physically challenged.

In this context, the Bill seeks to provide a climate governance framework that prioritises the needs of these groups in South Africa's adaptation and mitigation strategies. It envisages that this will be done through climate change approaches that are implemented in the context of sustainable development and, by implication, its ecological component, i.e., ecologically sustainable development. In addition to the above strategies for climate governance, South Africa's Presidential Climate Commission has recommended "fostering community stewardship of natural resources and considering indigenous knowledge" (Presidential Climate Commission, 2022, p. 20). This illustrates the importance of putting indigenous peoples first in formulating climate change mitigation and adaptation strategies.

4.4 *Batho Pele (People First) and the Judiciary*

There are several judgements in which South African courts have dealt with the application of *batho pele* principles, although the bulk of them do not deal with climate governance and environmental law. This is despite the fact that climate change litigation is increasing in South Africa. Some of the recent cases are spearheaded by indigenous peoples who feel a sense of responsibility for the conservation of the environment for the benefit of their communities and humanity. In the *Sustaining the Wild Coast* case, the court said that for these indigenous peoples, 'not only does the land belong to them, they also belong to the land.' Most of the cases emanate from government decisions to authorise oil and gas exploration on land occupied by indigenous peoples without obtaining their consent. While exploration and extraction of fossil fuels bring some benefits for governments, such as revenue collection and, to some extent, employment of local populations, it also brings pollution, which destroys natural habitats for land and marine animals, and fuels climate change (Buzmakov & Khotyanovskaya, 2020).

The recent case of *Sustaining the Wild Coast*, decided by the High Court of South Africa, is a classic example of a court putting the people first ahead of the interests of big oil and gas. In this case, the applicants sought an urgent interdict to stop Shell and its affiliates from continuing with seismic blasting off the Wild Coast for hydrocarbons. They contended that not only did Shell fail to do a proper consultation but also that its activities posed a grave and imminent danger to marine life. They supported their case with a body of scientific evidence that pointed out that seismic blasting emits massive sounds that affect humpback whales, dolphins, fish and other marine organisms, causing them to migrate and affecting their feeding and breeding habits. As such, this case was about securing ecologically sustainable development in the face of encroachment by a 'carbon major'—Shell.

The court found that the applicant communities were lawful holders of customary fishing rights and that they used the sea to practice their culture and beliefs. As such, they stood to be negatively affected by seismic blasting. This entitled them to judicial relief, as the courts have a duty to step in and protect customary practices and beliefs in areas where commercial activities would disturb the environment in which those customary practices are performed. The court also took issue with the fact that the stakeholder analysis developed by Shell had not identified the many small-scale and subsistence fishing communities along the coast where Shell was going to perform the seismic survey. Since Shell was legally obliged to consult the communities which were going to be impacted by seismic blasting, Shell had failed to satisfy the statutory requirements for an adequate consultative process, rendering it impossible for affected communities to grant informed consent to the seismic survey.

Also, the court found that the applicants had a right to be consulted before the granting of the exploration licence that authorised seismic blasting. This was not done, particularly because of the fundamental flaws underlying the consultation process in which traditional communities were excluded from the list of people who were consulted before the approval of the environmental management programme. In fact, people in the villages did not know that there was a consultation process underway. The notices outlining the application for an exploration licence by Shell were published in newspapers in a way that was designed to exclude the local traditional communities. This is because they were published in both English and Afrikaans, to the exclusion of isiZulu and isiXhosa, which are spoken by the affected communities. Hence, the consultation process was inadequate. Also, traditional communities were excluded in that after interested persons commented on the draft environmental management programme, notices were only sent to them directly, resulting in the further exclusion of traditional communities.

The court also found that the consultation process was invalid and wholly inadequate in that Shell only spoke to the traditional leaders and did not deem it necessary to speak directly to the local people. This was a contradiction of consensus-seeking in customary law, as it is established that kings and queens have no legal and customary authority to make representations and enter agreements on behalf of their traditional communities without the knowledge and consent of the traditional communities. Doing so means that the traditional communities would not have given informed consent. To make matters worse for Shell, it had consulted kings who did not have jurisdiction in the Wild Coast. As such, these kings did not have the authority to speak on behalf of customary fishers along a portion of the Wild Coast.

For the above reasons, the court interdicted Shell and its affiliates from continuing with the seismic blasting, costing them more than R1 billion (slightly more than £50 million) in total losses. The *Sustaining the Wild Coast* judgement reveals that putting people is fundamental to the granting of environmental authorisations for fossil exploration, even if it disadvantages big oil and gas. While it may be true that public participation is a procedural requirement that is capable of independent enforcement without reference to *batho pele*, the reality is that *batho pele* is more than about ticking boxes to secure environmental authorisation: it is about genuinely putting the people first, hearing them out and safeguarding their interests. Shell had to lose the case because it did not place the needs of customary communities, who stood to be affected, at the forefront and was not transparent with them. It placed its potential revenue gains at the forefront with little to no regard for the consequences of its actions on the Wild Coast people. Thus, its actions can be summarised as money first, not people first.

5 Conclusion

This chapter explored the potential of indigenous values in climate change mitigation and adaptation through ecologically sustainable development in South Africa. In argues that in this regard, *batho pele*, the South African indigenous value of putting people first, is an essential component of balancing the competing interests of protecting ecological integrity and advancing economic and social development in fulfilment of the constitutional obligation to secure ecologically sustainable development. Our analysis is motivated by the fact that despite their contribution to ecological integrity through indigenous sustainability, indigenous peoples are more exposed to the negative consequences of climate change, pollution and the destruction of biodiversity and the environment. Hence, indigenous peoples have a huge interest in putting their knowledge systems and values to work to protect the climate and preserve the environment and ecological integrity. We are also motivated by that whereas scholars recognise the importance of indigenous knowledge systems and values in South Africa's environmental governance framework, their discussions leave a gap on other indigenous values which may also contribute to ecologically sustainable development, such as *batho pele*.

In this chapter, we have tried to show that implementing principles that underlie *batho pele* could help to deliver on the constitutional obligation to secure ecologically sustainable development. This is because *batho pele* is already embedded in South Africa's environmental policy and regulatory approaches. Its definitional elements hold a strong promise for securing ecologically sustainable development. Implementing the ideals that underlie *batho pele* will ensure that executive decisions that impinge on the climate and the environment, such as the granting of exploration licences for oil and gas, serve the best interests of indigenous peoples and that it enables them to enjoy constitutional rights such as the right to have the environment protected through securing ecologically sustainable development.

Our approach to infusing *batho pele* in efforts to secure ecologically sustainable development also emanates from our acceptance of the presence of fault lines in the sustainable development discourse, particularly the issue that present conceptions of sustainable development are not optimised to advance the interests of indigenous peoples in preserving ecological integrity but that they serve the materialistic approaches of global capital. We argued that the consideration of indigenous values such as *batho pele* would ensure that the interests of indigenous communities are genuinely considered and prioritised in the making of environmental decisions that affect them. In our view, indigenous values may be the ideal premise on which to develop a truly South African jurisprudence for the achievement of human dignity, equality and freedom, even in the most complex areas of law like climate change and environmental law. As a result, we hope that *batho pele*, which has been neglected in the discourse, will gain better recognition as an indigenous perspective from which to secure ecologically sustainable development.

References

Addaney, M., & Cobbinah, P. B. (2019). Climate change, urban planning and sustainable development in Africa: The difference worth appreciating. In P. B. Cobbinah & M. Addaney (Eds.), *The geography of climate change adaptation in urban Africa* (pp. 3–26). Palgrave Macmillan.

Adelman, S. (2013). Rio+20: Sustainable injustice in a time of crisis. *Journal of Human Rights and the Environment, 4*(1), 6–13. https://doi.org/10.4337/jhre.2013.01.01

Baleni v Regional Manager: Eastern Cape Department of Mineral Resources [2020] 4 All SA 374 (GP).

Bosselmann, K., & Kim, R. E. (2013). International environmental law in the Anthropocene: Towards a purposive system of multilateral environmental agreements. *Transnational Environmental Law, 2*(2), 285–309. https://doi.org/10.1017/S2047102513000149

BP Southern Africa (Pty) Ltd v MEC for Agriculture, Conservation, Environment & Land Affairs [2004] 3 All SA 201 (W).

Buzmakov, S. A., & Khotyanovskaya, Y. V. (2020). Degradation and pollution of lands under the influence of oil resources exploitation. *Applied Geochemistry, 113*, 104443. https://doi.org/10.1016/j.apgeochem.2019.104443

Carruthers, D. (2001). From opposition to orthodoxy: The remaking of sustainable development. *Journal of Third World Studies, 18*(2), 93–112. https://www.jstor.org/stable/45193956

Chikuruwo, R. E. (2022). *Ecologically sustainable development as a constitutional value: South African perspectives.* [LLD thesis, NWU], Potchefstroom. http://hdl.handle.net/10394/39520

Church, J. (2012). Sustainable development and the culture of ubuntu. *De Jure, 45*(3), 511–531. http://ref.scielo.org/xprqcy

Climate Change Bill [B 9—2022].

Constitution of the Republic of South Africa, 1996.

Department of Environmental Affairs. (2008). *Strategic plan for the environmental sector 2009–2014*. https://www.gov.za/sites/default/files/gcis_document/201409/200914strategicplan0.pdf

Department of Public Service and Administration. (1997). *White paper on transforming public service delivery*. https://www.gov.za/sites/default/files/gcis_document/201409/183401.pdf

Fuel Retailers Association of Southern Africa v Director-General: Environmental Management, Department of Agriculture, Conservation and Environment, Mpumalanga Province 2007 10 BCLR 1059 (CC).

Government of the Republic of South Africa. (2011). National climate change response white paper. https://www.sanbi.org/documents/national-climate-change-response-white-paper/

Government of the Republic of South Africa. (2017). National Climate Change Adaptation Strategy Republic of South Africa. https://www.dffe.gov.za/sites/default/files/legislations/session2_draftnational_adaptationstrategy.pdf

Jerez, M. M. (2021). Challenges and opportunities for indigenous peoples' sustainability. https://www.un.org/development/desa/dspd/2021/04/indigenous-peoples-sustainability/

Kothari, A., Demaria, F., & Acosta, A. (2014). Buen vivir, degrowth and ecological swaraj: Alternatives to sustainable development and the green economy. *Development, 57*(3), 362–375. https://doi.org/10.1057/dev.2015.24

Kotzé, L. J. (2016). *Global environmental constitutionalism in the Anthropocene*. Hart Publishing.

Marten, G. G. (2001). *Human ecology: Basic concepts for sustainable development*. Earthscan.

McGregor, D., Whitaker, S., & Sritharan, M. (2020). Indigenous environmental justice and sustainability. *Current Opinion in Environmental Sustainability, 43*, 35–40. https://doi.org/10.1016/j.cosust.2020.01.007

Mec: Department of Agriculture, Conservation and Environment v Htf Developers (Pty) Ltd 2008 4 BCLR 417 (CC).

Merafong Demarcation Forum v President of the Republic of South Africa 2008 5 SA 171 (CC).

Mwanza, R. (2019). Harnessing the transformative potential of the constitutional human right to a clean and healthy environment in the context of corporate environmental damage in Kenya: A critical perspective. *Journal of Human Rights and the Environment, 10*(2), 215–238. https://doi.org/10.4337/jhre.2019.02.04

Mwanza, R. (2022). Framing the normative role of the right to a healthy environment: Thinking with internormativity, embodiment and emergence. *Journal of Human Rights and the Environment, 13*(2), 349–370. https://doi.org/10.4337/jhre.2022.02.02

National Environmental Management Act 107 of 1998.

Presidential Climate Commission. (2022). A framework for a just transition in South Africa. https://pccommissionflow.imgix.net/uploads/images/A-Just-Transition-Framework-for-South-Africa-2022.pdf

S v Makwanyane 1995 6 BCLR 665 (CC).

South Africa. (2021). First nationally determined contribution under the Paris Agreement. https://unfccc.int/sites/default/files/NDC/2022-06/South%20Africa%20updated%20first%20NDC%20September%202021.pdf

Springer, S. (2016). Fuck neoliberalism. *ACME: An International Journal for Critical Geographies, 15*(2), 285–292.

Sustaining the Wild Coast NPC v Minister of Mineral Resources and Energy [2022] 1 All SA 796 (ECG).

Tsosie, R. (2019). Indigenous sustainability and resilience to climate extremes: Traditional knowledge and the systems of survival. *Connecticut Law Review, 51*, 1009. https://opencommons.uconn.edu/law_review/407

United Nations Climate Change. (2021). South Africa. https://climatepromise.undp.org/what-we-do/where-we-work/south-africa#:~:text=South%20Africa%27s%20updated%20mitigation%20targets,communicated%20in%20the%20first%20NDC.

Van Der Merwe v Taylor 2007 11 BCLR 1167 (CC).

Van Norren, D. E. (2020). The sustainable development goals viewed through gross national happiness, ubuntu, and buen vivir. *International Environmental Agreements: Politics, Law and Economics, 20*, 431–458. https://doi.org/10.1007/s10784-020-09487-3

Viñuales, J. E. (2013). The rise and fall of sustainable development. *Review of European, Comparative & International Environmental Law, 22*(1), 3–13. https://doi.org/10.1111/reel.12021

Index

M. Addaney et al. (eds.), *Climate Change in Africa*,
https://doi.org/10.1007/978-3-031-30050-9

B

D

FSC
www.fsc.org
MIX
Papier | Fördert
gute Waldnutzung
FSC® C083411

Kolibri 360 GmbH
Ferdinand-Jühlke-Straße 7
99095 Erfurt, Deutschland
produktsicherheit@kolibri360.de